Vaughan Cornish

Short Studies in Physical Science

Mineralogy, chemistry, and physics

Vaughan Cornish

Short Studies in Physical Science
Mineralogy, chemistry, and physics

ISBN/EAN: 9783337034788

Printed in Europe, USA, Canada, Australia, Japan

Cover: Foto ©berggeist007 / pixelio.de

More available books at **www.hansebooks.com**

SHORT STUDIES

IN

PHYSICAL SCIENCE

MINERALOGY, CHEMISTRY, AND PHYSICS.

BY

VAUGHAN CORNISH, M.Sc.,

ASSOCIATE OF THE OWENS COLLEGE;

AUTHOR OF "PRACTICAL PROOFS OF CHEMICAL LAWS."

ILLUSTRATED.

LONDON:

SAMPSON LOW, MARSTON & COMPANY

LIMITED,

St. Dunstan's House,

FETTER LANE, FLEET STREET, E.C.

1897.

LONDON :
PRINTED BY WILLIAM CLOWES AND SONS, LIMITED,
STAMFORD STREET AND CHARING CROSS.

PREFACE.

SEVERAL of these Studies have appeared in *Knowledge.* One or two are reprinted from the *Speaker*, by the kind permission of the editor. The rest are new.

VAUGHAN CORNISH.

BRANKSOME CLIFF,
BOURNEMOUTH,
September, 1896.

CONTENTS.

MINERALOGY.

LIST OF ILLUSTRATIONS.

MINERALOGY.

SHORT STUDIES

IN

PHYSICAL SCIENCE.

CHAPTER I.

CRYSTALS.

SECT. I.—THE CHARACTER OF CRYSTALS.

IN the fresh-fallen snowflake skeleton crystals are readily
seen, and the crystalline character can often be distinguished
in ice; while in the road on a frosty morning the characteristic angle of ice crystallisation, the angle of the equilateral
triangle, 60°, can be seen where the mud has been hardened
in the ruts, and long spikes of ice trace a complex pattern on
the frozen ground. The alum-coated basket and the glistening
branches of the "lead-tree," formed when a piece of zinc is
placed in a solution of sugar of lead, are familiar to those who
played as children with the playthings of science. A few drops
of sal-ammoniac spread in a thin film on a slip of glass, show
the process of crystallisation beautifully under the microscope,
long spikes and branches shooting out with marvellous
rapidity as the moisture exhales, ranging themselves all the
while with geometrical precision. A drop of ink placed on
a microscope slide deposits crystals on evaporation, which
are well seen with a moderate magnifying power.

In the brown granite slabs and pillars, commonly used in buildings of a certain pretension, the sections of the large brown crystals of feldspar are noticeable from the regularity of their angles and the peculiar lines symmetrically traced upon their polished surface. Diamonds, of course, are crystals, and so are the sapphire, ruby, and emerald ; but the natural crystal faces are replaced in the cut stones by the lapidary's facets, and the gems seem as much the product of art as of nature.

In the case of the diamond, the ruby, and the sapphire, the jeweller's art is needed to bring out the latent beauties of the stone. The diamond, as found in the mine, looks rather like a lump of white wax ; the ruby and sapphire are generally found as rough and water-worn pebbles. With all respect to Mrs. Hemans, it must be in a " Better Land,"

> " Where the burning rays of the ruby shine,
> And the diamond lights up the secret mine."

At all events, in South Africa and Burmah one does not find these gems ready cut. Perhaps the song refers to the country of Flap-doodle, where, according to Charles Kingsley, the little pigs run about ready roasted. This, however, is a digression, *revenons à nos moutons !*

The application of the jeweller's art is not very satisfactory in the case of the emerald, for the cut facets do not compare to advantage with the bold angles of the natural crystals, whose regular six-sided prism, with smooth and lustrous faces, often makes a splendid object projecting from the matrix in which it is found.

We may refer the reader to the specimens in the Mineral Gallery of the Natural History Museum, Cromwell Road, South Kensington, where some fine examples are shown in Case 29c.

Crystals, cut and natural, afford the brightest spots of beauty in the inorganic world, excepting only a dewdrop in the sunlight, the stars in a frosty sky, and, last and best of all, the unnumbered flashings of the sunlight on a rippling summer sea.

If the asthetics of crystals were more studied, people would, perhaps, relinquish the practice of piling up crystallised minerals in a miniature rockery under a glass case.

The *growth* of a crystal reminds one of the growth of a living thing. The artificial production and growth of minerals and gems is a difficult matter to achieve, but many substances which crystallise out from their solution in water will readily yield crystals as large as those occurring in rocks and veins, although their softness and solubility prevent them from taking a high polish, or standing wear and tear. Numerous specimens are shown at the Natural History Museum.

In growing crystals, the object kept in view is the symmetrical development of a single individual, not the accumulation of a confused aggregate like that of the alum-basket or lead-tree.

A strong solution of the substance, say, for instance, copper sulphate, nickel-ammonium sulphate, or alum, is left to evaporate in a suitable vessel; flat-bottomed glass "crystallising dishes" are the best. The dish is carefully protected from dust, and after a time, a deposit of small crystals forms on the bottom. The largest and most evenly developed is selected, and is placed on the bottom of a dish containing a solution of the substance, a muslin bag containing a supply of the material being suspended in the upper layers of the solution to keep it saturated. The crystal is from time to time turned over, so that each face may have equal opportunities of receiving material for growth. *On this condition only will the development be symmetrical.* Great care and watchfulness are needed to secure a regular growth. Sometimes the crystal is suspended in the solution by a fine thread, so that the solution may have simultaneous access to every face.

The fact that the regular growth of a crystal is only brought about by special conditions, indicates that the *shape* of a crystal is to a great extent a non-essential character.

The crystal models, showing the "ideal" crystalline forms, are in this way misleading to the learner, for they make the non-essential character of even development appear to be the essence of the matter.

There are only three *essential* characters of crystals. Firstly, they are bounded by *planes*, the crystal faces. This character is, of course, reproduced in the models. Secondly, the crystal faces have definite, constant, inclinations to one another. These crystal *angles* (by which we mean the inclination of two *faces*, not the points or solid angles) are, of course, reproduced in the models of crystals. The third essential character of crystals, that which chiefly gives them their great natural and scientific importance, is their internal structure—the *crystalline* structure. A substance may be perfectly crystalline, and yet show no crystal faces or external symmetry. For example, the quartz of granite, which was the last portion of the rock to solidify, has filled whatever spaces remained for it to fill, and is altogether irregular in its distribution, having no proper shape or form. Nevertheless it is perfectly crystalline.

Every property of a crystalline body, elasticity, conducting power for heat and electricity, and the action upon light, shows that every minutest portion is built upon a systematic plan.

The properties in all crystalline bodies are the same round any one point as round any other point. The properties are not necessarily or generally the same in each *direction* round any point. Take the fundamental property of elasticity, the elasticity of shape, which is the especial peculiarity of solids. Crystalline bodies may have three different elasticities in different directions, called the axes of elasticity; or the elasticity may be the same in two directions and different in the third. Finally the *isotropic* crystals have the same elasticity in the direction of all three axes.

Isotropic crystals—alum, for instance—most nearly resemble coherent amorphous bodies, such as glass, which are also isotropic, *i.e.* have the same properties in all directions. The difference between them may, perhaps, be expressed by the saying that uniformity in a non-crystalline substance is like the uniformity in a vast crowd. The number of individual parts is very great in proportion to the possible differences between different individuals. Any portion of a crowd containing a large number of individuals appears practically the same as any other such portion. The smallest parts of a solid body with which we can deal, undoubtedly contain vast numbers of individual particles. In an isotropic crystal, the regularity in all directions is more than apparent. It is real, like the regularity in a battalion of infantry. The isotropic crystal differs from the non-crystalline body by "*regimentation.*"

The crystal is the true characteristic solid, not the exception among solids. Glass and wrought iron are little more than stiffened liquids. This is well seen if we study the action of heat upon them. There is no definite temperature at which glass and wrought iron cease to be solid and become liquid; the change is gradual and progressive, passing as the temperature rises through all stages of plasticity. Crystals, on the other hand, have a proper melting point. The transformation to the liquid state is abrupt, as is the change from gas to liquid below the critical temperature of the gas, which is marked by the sharp surface of separation in the containing vessel. Above the critical temperature, there is no liquid surface, as can be shown by the familiar experiment with carbonic acid gas in a sealed tube. Above the critical temperature, there is no proper distinction between liquid and gas, or, as it may also be expressed, there is no properly liquid state. This corresponds roughly to the condition of glass, which is never properly a solid, though very different from an ordinary liquid.

In non-isotropic crystals, such, for instance, as ruby and emerald among minerals, and copper sulphate and prussiate of potash among soluble salts, the elasticity and other properties are different in different directions. In

this they more nearly resemble timber than glass or
iron.

The most striking manifestation of the different elasticities
of a non-isotropic crystal is afforded by its action upon a
ray of light. A ray of light consists of vibratory motion
propagated after the manner of a wave, the vibration at one
point setting up a vibration at the next point, and so on;
so that the disturbance is propagated although the disturbed
parts are not themselves translated. When a sound wave
travels in air, the disturbed particles of air travel a *short*
distance forwards and backwards—the vibrations of the air
particles being in the direction in which the wave travels.
In the case of light, on the other hand, the vibrations take
place at right angles to the direction of the ray.

A surface drawn through the positions to which a
luminous disturbance, or light wave, has reached at any
moment is the *wave-front* at that moment. Suppose the
source of light be far distant—the sun, for instance—and
that we are dealing with so much of the luminous disturbance
as will reach a body of moderate size, a "ray" of light, in
fact. The ray is supposed to be traversing a vacuum or
any isotropic body, whether a liquid or an isotropic solid.
The wave-front of the ray of light is a flat surface at right
angles to the direction of the ray. The vibrations may
take place in any direction at right angles to the ray.
When the ray of light comes direct from a hot body,
the sun, or a gas-flame, or a heated platinum wire—the
direction of vibration in a plane at right angles to the ray
changes from moment to moment. The most general form
of the vibration is an ellipse, lying of course in a plane at
right angles to the ray of light, but the shape of the elliptical
vibration in this fixed plane where we have cut the ray
constantly changes. Sometimes the two axes are equal;
the vibration is then circular. Sometimes one axis becomes
zero, when the vibration is rectilinear. In a very short
space of time every phase—straight line, narrow ellipse,
wide ellipse, circle, and so on back to straight line—is

repeated, but with ever-varying inclination of the straight line, or of the axis of the ellipse.

Now, suppose a slice from a transparent non-isotropic crystal to be interposed in the path of the ray. The vibrations are at once reduced to order by the ordered structure of the crystal. The elasticity of the crystal is different in different directions. In certain directions, but not in others, a displacement will produce a force of restitution along the direction of the displacement. Only in such direction can vibration be set up, and the light-movement in the wave-front is constrained to a vibration in a straight line. This is the simplest form of *polarised* light, and is the usual mode of propagation of light within a non-isotropic crystal; it may be termed *rectilinear* polarised light.

Two such waves of light may be simultaneously propagated within a crystal with different velocities depending upon the different elasticity of the crystal in different directions.

When the light emerges from the other side of a slice of crystal, and continues its course in air, or other isotropic body, there is only one possible velocity of propagation. The two waves which travelled independently in the crystal may now compound and coalesce. When the two vibrations, at the moment of leaving the crystal, compound into one, its general form is an ellipse, but in a special case may be a circle, *i.e.* an ellipse with equal axes. One wave, at the moment of leaving the crystal, is always the same amount behind the other, and since the directions of vibration *in* the crystals are fixed, the ellipse formed by their coalescence on leaving the crystal has a fixed shape, and the directions of its axes are fixed. The light which travels on through the air after leaving the crystal differs therefore from ordinary light. The vibrations in any plane cutting the ray at right angles are constant instead of perpetually varying. This is called *elliptically polarised* light. The limiting case is called *circular* polarisation.

Turning from the *structure* to the *forms* of crystals, we find in crystals the morphology of inorganic solids.   Unlike most forms of vegetable and animal growth, crystals are bounded by flat surfaces.   The crystal faces are not placed in a haphazard manner according to circumstances.   However much the conditions of formation and growth of a crystal may modify the development of particular faces, or may cause certain faces to appear on one specimen which are not present on other specimens of the same substance, there is no deviation from the following law : The form of a crystal of any substance may be referred to a system of three axes whose angles and relative lengths have fixed values for each substance.   The different faces of the crystal, if moved parallel to themselves to cut one axis in a given point, cut the other axes *so that the segments of either of these stand in simple ratios to each other* (Ostwald).

One of the consequences of this "law of rationality of the indices" is that the angles between the faces of a crystal do not vary—as do the sizes and shapes of the faces—with the conditions of growth.

The observed forms of crystals have been classified according to their degrees of symmetry into six "systems of crystallisation." It has been demonstrated mathematically that the six observed systems account for all the different degrees of symmetry which are possible under the law of rationality of indices.

Attempts have been made of late years to discover the actual arrangement of molecules in crystals by arranging points in the various ways which will satisfy their observed symmetry.   The arrangement of points must fulfil the requirement that the properties of a crystal are the same round any one point as round any other point, and must be in accordance with the law of " rationality " of indices. Bravais and Sohnke have made such progress in working out the possible arrangements of crystal particles that sanguine people hope to find a basis for the comparison of the representations of crystal structure with molecular structure as represented in the graphic formulæ of chemistry.

Crystal forms present some curious analogies to the animal world, analogies that must not be pressed too far, but which are interesting in themselves, and which may serve to enlist the interest of the students of botany and zoology.

Right and left handedness is one of the most important modifications of crystal symmetry. Half the possible faces of a crystal form being omitted, we have either the right-handed or the left-handed half form, or hemihedral form. Two half forms growing together sometimes give the curious re-entrant angle seen, *e.g.*, in the arrow-headed form of gypsum. These are called twin crystals. Some twins are composed of two right-handed forms or two left-handed forms, others of a right-hand and left-hand form.

The whorls of univalve shells afford another instance of right and left handedness in Nature, the whorls in some species being right-handed and in others left-handed. Perhaps it is hardly necessary to state that the right and left hand differ by not being superposable, though otherwise similar. The relation of a right-hand form to a left-hand form is that of an object to its image in a mirror.

Sometimes a crystal is found with an external form which belongs to another species of crystal. This is called pseudomorphism, and presents an interesting analogy to mimicry among animals. It is met with, *e.g.*, when a crystal of calcite has been dissolved from a cavity in which it has left a cast of its shape, and quartz has subsequently crystallised in the cavity. The quartz has, from the necessity of the case, accommodated itself to the outward form of the calcite. As in the case of mimicry, so in pseudomorphism, the resemblance to the foreign form is a striking one, but not carried out in structural detail.

Sometimes, by peculiarly complex arrangements, crystals simulate in their external form a higher symmetry than is warranted by their real structure, a habit to which it is not difficult to find a parallel in the practice of the highest of all animals.

When crystals of various sizes are present together in such circumstances that they struggle for existence (*e.g.* when alternately heated and cooled in the presence of a solvent), the large crystals grow at the expense of the small ones, which present a relatively larger surface to the action of the solvent. When the solvent on cooling yields up the dissolved matter, it deposits on the nucleus afforded by the remaining larger crystals. The large crystals eat up the small ones.

In many organisms which contain a large quantity of calcerous matter, the processes of crystalline and of animal growth go on side by side, giving rise to a variety of interesting phenomena, some of which we proceed to describe.

The description of crystallisation in shells will involve the repetition of part of what has been already dealt with ; but repetition is not altogether without advantage in dealing with this intricate subject.

SECT. II.—CRYSTALLINE GROWTH IN SHELLS.

If different materials have the same chemical constitution they are generally regarded by the chemist as being essentially identical, and are looked upon as *varieties* of the same substance. Carbonate of lime is a familiar example ; chalk, limestone, marble, Iceland spar, and aragonite are all composed of carbonate of lime, and are spoken of as different varieties of carbonate of lime.

This does not, however, represent the view of the mineralogist. Chemical composition is only one among several *criteria* considered in the definition of a mineral species. It often happens that minerals differing fundamentally in their crystalline form, and in the internal structure which is connected with the external form, have the same chemical composition. These are regarded by the mineralogist as distinct species, notwithstanding the identity of chemical composition. Fundamental difference of crystalline form is the basis of differentiation among minerals of which the composition is identical.

We will endeavour to make clear the principles on which it is decided whether a difference of form is to be regarded as fundamental. In mineralogical collections, ranged side by side with the well-known rhomb of Iceland spar, may be seen a vast variety of crystalline forms of carbonate of lime. In general appearance they differ greatly from one another ; yet the crystallographer will tell one that there is no *fundamental* difference in these forms, which are all intimately related to that of the rhombohedron. This relationship is a matter of geometry, which we must be content to state merely in a general manner. A crystal is a body bounded by certain plane surfaces—the faces of the crystal. The arrangement of a system of planes is best understood by considering how their position is related to three lines intersecting at a point, these lines being termed *axes*. A crystalline form is defined, first, by the position of these ines ; secondly, by the manner in which the position of the planes is related to that of the lines. All the various forms which, in the mineralogical collection, are placed in proximity to the rhomb of Iceland spar, are forms which can be built up on the same system of axes, and the positions of the planes or faces, with respect to these axes, are all related to one another according to a simple geometrical law. All these specimens are reckoned as belonging to one mineral species, to which the name *calcite* is given. (See Case 18E at the Natural History Museum in Cromwell Road.)

In another compartment of the mineralogical collection will be found another set of crystals, also composed of carbonate of lime. These, however, are members of another mineral species, known as *aragonite*. (See Case 17A at the Natural History Museum.)   Their forms are related among themselves by a simple geometrical law, but the plan of construction is radically different from that of the calcite forms.   The system of planes representing the faces of the crystals cannot be built up on the same three axes as those of the calcite forms, and the law connecting the positions of the planes themselves is different in the two cases.

For the mineralogist, carbonate of lime comprises two mineral species — calcite and aragonite — and all the different varieties are classed under one of these two names.

The question may be asked—Is there not something fanciful in basing a classification of material substances merely on certain abstract ideas of symmetry?   Calcite and aragonite are composed of the same kinds of stuff or matter; if our minds were not gifted or encumbered with abstract notions about symmetry, should we discover any difference of properties between these two mineralogical species?

As a matter of fact, the properties of crystalline substances furnish a striking example of the real and intimate relation of our ideas of symmetry with the actual constitution and properties of matter.   Every property of a crystalline substance is related to the particular symmetry of its form.   Thus, take the case of the action of the substance upon light.   A ray of light is affected in the same way by its passage through any crystal of calcite ; all specimens of aragonite behave alike in their action on light ; but the mode of action is entirely different in the case of aragonite from that of calcite.   For exhibiting these differences of behaviour in the most distinct manner, an elaborate instrument is employed, the stauroscope, in which *convergent polarised* light is used.   We cannot enter here into the method of using the instrument, but in most textbooks of mineralogy will be found plates showing the different *interference* phenomena afforded by a crystal of calcite and one of aragonite.   A fragment of a crystal shows the phenomena characteristic of its species, which do not depend upon the specimen possessing crystalline faces. The action upon light depends upon the internal structure of the material of which the crystal is composed, which, as the above example shows, is intimately related with the crystalline form.

We see, then, that the physical properties justify the mineralogist in sorting the varied crystalline forms of carbonate of lime into two classes, and in characterising every member of each class by one of two names—the names of two mineral species.

We have said that *all* varieties of carbonate of lime are classed as calcite or aragonite, and many of these varieties do not show crystalline form at all, as, for instance, limestone and marble. What data justify the mineralogist in calling limestone and marble "calcite"? We have said that where crystalline symmetry differs, all physical properties are different. One of the most important, and one of the most nearly constant properties of each kind of matter, is its specific gravity. Every specimen of calcite has a specific gravity of 2·72, or not differing from 2·72 by more than one or two units in the second decimal place. Similarly, all specimens of aragonite have a specific gravity of very nearly 2·93. In limestone and marble we have the so-called *massive* carbonate of lime, the material not having had the opportunity to assume the crystalline form. How are we to determine whether limestone is calcite, aragonite, or yet another mineral species? The specific gravity of limestone is found to be that of calcite, viz. about 2·72, as is also the case with marble. We conclude, therefore, that the carbonate of lime in limestone and marble is calcite.

It has long been a familiar fact that certain organisms have the power of secreting carbonate of lime from solution to build up the hard portion of their shells. The shell of the oyster, for instance, of the crab, and of the common whelk, are complex structures of organic matter and carbonate of lime, and in the fossilised remains of shells the general form of the shell is preserved by the carbonate of lime after the decomposition of the organic matter.

Several questions of interest are suggested by this case of combined mineral and animal growth. Does the carbonate of lime in calcareous organisms exist in some special modification different from those known in the inorganic world? How far is the mode of growth of the animal tissues constrained by rigid laws of crystallisation? Or, on the other hand, do the forces brought into play in animal growth mask, or even overpower, the operation of the ordinary process of the crystallisation of a substance from solution?

The answers to these questions are furnished by the investigations of Dr. Sorby and others, which have been published at intervals during the last sixteen years.

In the first place, it has been shown that in calcareous organisms we have *not* to deal with any new species of carbonate of lime. In every case the optical properties and the specific gravity show that the shells contain either calcite or aragonite. Some animal species secrete carbonate of lime as calcite, others as aragonite.

The mode of growth of the shell is in general a compromise between the mineral and the organic, in some cases the influence of the first factor having the predominance, in others that of the second.

The direction of the *principal axis* of the crystal is always related in a definite manner to the surface of growth of the shell—the symmetrical arrangements which result from this relation producing very beautiful appearances when sections are examined under the microscope. Thus, the inner shell of *sepia* (the cuttle-fish) shows innumerable crystals of aragonite ranged in parallel rows, whilst the mineral portion of the spines of *echinoderms* consists of a single crystal of calcite greatly developed in one direction.

Some organisms, as we have said, secrete or produce calcite, others aragonite. Other cases, again, are known in which one portion of the shell is built up of, say, aragonite, and during the subsequent growth of the organism its habits or powers undergo a change, the rest of the shell being built up of calcite.

It will be seen that we are dealing with a mixed study, at once mineralogical and biological. It is peculiarly interesting to find that the influence of the laws of evolution is apparent even when studying the mineralogical aspect of the subject. It is well known that the embryo shows the past history of the species, the development of the individual furnishing an epitome of the history of development in the race. Now, the shell of the common whelk, which may be found on the sea-beach, is composed of calcite, except a small portion which is that *first* formed in the growth of the organism. The examination of the *fossil* species of the whelk tribe shows that they are composed wholly of aragonite, the composition of the whole shell of the early individuals of the race being identical with that of the embryo of their modern representative.

It was long since observed that in fossiliferous beds which are permeable to water, certain calcareous shells are preserved, whilst others are only represented by their impressions. This is the more remarkable, because some of the most massive shells have disappeared, whilst others of delicate structure remain.

The form of the impression enables the species of the shell to be identified, and it is found that the shells which have been removed are those of species known to secrete their carbonate of lime as aragonite. It was known that under many conditions aragonite is less stable than calcite, and it was assumed from the above observations that aragonite is more readily dissolved than calcite by water containing carbonic acid. A few years ago this point was made the subject of experimental investigation, with somewhat singular results. (See Vaughan Cornish and Percy F. Kendall, "On Calcareous Organisms," *Geol. Mag.*, Feb, 1888.)

Pure and well crystallised specimens of calcite and aragonite were subjected in a powdered state to the action of a solution of carbonic acid under similar conditions. No difference of solubility was detected.

Powdered calcite and aragonite fossil shells gave a like result.

It was found, however, when the complete shells were suspended in a solution of carbonic acid, that those of aragonite were much more readily acted upon than those of calcite, and, further, that the coherence of the aragonite shells was soon destroyed, so that the slightest agitation of the water was sufficient to entirely disintegrate the shell, reducing it to the condition of a fine powder or mud.

The disappearance of aragonite fossils is explained by these experiments ; it is due not to the greater solubility of aragonite, but to a mode of structure of shells composed of aragonite which facilitates their solution and disintegration. The experiments gave somewhat unexpected results also in the case of calcite fossils. It was found that they were acted upon with considerable rapidity by the solution of carbonic acid, but that they retained their compactness, and even the delicate details of marking, after losing as much as 15 per cent. of their weight through the action of the solvent.

It is thus evident that the calcite fossils found in porous or permeable beds *simulate* an immunity from the action of carbonic acid, which they do not in reality enjoy. In such beds as these a large quantity of carbonate of lime goes into solution, and cavities are formed having the shape of the fossils which have been removed. It frequently happens that at a subsequent period, carbonate of lime crystallises out from solution in these cavities. When crystallisation takes place at the ordinary temperature, calcite is formed—a fact which is readily established by laboratory experiments. Consequently *casts* are formed in calcite of aragonite fossils. They are readily distinguished from the originals by their translucency, aragonite fossils being always opaque.

Thus the beautiful *ammonites* often found of a material resembling fine alabaster are replicas only, no trace of the original aragonite shell remaining. Many other such cases occur.

This subject belongs to a sort of no-man's land in science. The borderlands of the different sciences are apt to be somewhat neglected, but are capable of yielding a fruitful harvest of results.

SECT. III.—THE STUDY OF ROCKS.

There are two ways of writing a guide-book to a national collection. The first is to write something very simple, having regard to the needs of the average visitor. This is what Sir W. H. Flower has done in his description of the Introductory Collections which are placed in the central hall of the museum in Cromwell Road. The second method is to write a learned treatise adequate to the splendour of the collections themselves, which is the character of Mr. L. Fletcher's "Introduction to the Study of the Rocks." It is a valuable contribution to scientific literature, not an ordinary hand-book. Although not popular in the sense of being easy reading, it is nevertheless available to all who seriously desire to become acquainted with the subject, for there is no assumption of previous knowledge on the part of the reader. The author proceeds from first principles, and builds up the science of the rocks by a logical method, which renders the progress of the learner as sure and as laborious as that of the student of mathematics. The mathematical bent of the author's mind is, indeed, evident throughout the book, though, of course, the symbols of mathematics have no place in the treatment of his subject. For example, the definition of rock presents a logical rather than a practical difficulty, and Mr. Fletcher works out his definition with the loving care of a logician; but we expect the illustrative specimens will do more to teach the visitor what a rock *is* than all the logic in the world. Thus, the beautiful examples of Shap granite, in

one of the window-cases, show at a glance that any one
piece of rock is similar to any other piece of the same
rock, provided that the piece be large enough to include
the several mineral constituents of which the rock may be
composed.   It is one of many instances in which a material
can be regarded as homogeneous as long as we are not
dealing with its ultimate particles.

Perhaps the most generally interesting part of the book
is the discussion of the mode of formation of granite, and
of the relations between granite and other rocks (such as
rhyolite and obsidian) which have the same average chemical
composition.   Obsidian, a glass-like rock, is formed by the
sudden solidification of a composite liquid, of which the
components have no opportunity to separate into distinct
minerals.   In rhyolite, a rock less hastily produced from
the *same* kind of liquid magma, some of the mineral constituents have aggregated in the form of crystals which are
imbedded in a glassy ground mass.   Lastly, granite, formed
from the same kind of fluid magma, has slowly and steadily
solidified at great depths, where the masses of superincumbent material have rendered cooling slow, and have provided the constraining force of great pressure.   Under
these circumstances all the materials of the fluid magma
have assumed the crystalline state, and we have a rock,
called holo-crystalline, which is entirely composed of inter-
penetrating crystals of distinct mineral species.   The process
we have described would account, Mr. Fletcher says, " for
the fact that granite is generally found only among the
oldest rocks; for the occurrence of granite at the surface of
the earth's crust would only take place when denudation
has removed the enormous mass of superincumbent material
to which the enormous pressure was directly or in-
directly due, and therewith also the lava-streams which had
their source in the now granite reservoir."   It will be
noticed that this does not in any way imply that granite is
not *formed* at the present time.   It may be in actual
formation at this very moment deep down in the earth's

crust, but not till long ages hence will nineteenth-century granite be exposed at the earth's surface.

Mr. Fletcher adds some new words to the vocabulary of science—a doubtful boon, were it not that each new word docks two syllables from an old one, so that there is a balance on the right side. "Petrological," for instance, is reduced to "petrical," and "lithological" to "lithical." Undoubtedly the keeper of the national rocks and stones is the proper person to have initiated these changes of terminology, which, moreover, seem to be in themselves desirable.

The specimens introductory to the study of the rocks are set out in the window-cases on the side of the mineral gallery to the left of the entrance. Here are examples illustrating types of rock, modes of origin of rocks, joint planes, faulting, cleavage, cementing material, weathering, and so forth. The learner should also examine the window-cases I. to V. on the same side of the gallery, which contain the admirable series of specimens introductory to the study of mineralogy, the science of which petrology is a part. The introductory specimens contained in these cases, with their explanatory labels, afford, we believe, the readiest means of obtaining a general idea of the scope of this fascinating but difficult science. Mere reading is not sufficient, and "field work" is difficult to carry out in England, especially in the Southern counties.

We have sometimes regretted that the cases introductory to the study of minerals have not been removed to the central hall, now that the other departments of the Museum have been provided there with a similar introduction by way of typical specimens and explanatory labels. The entire absence of minerals from the central hall has the undesirable effect of emphasising the separation of the studies of organic and inorganic natural history. On the other hand, it must be admitted that the specimens can be better seen in the excellently lighted window-cases of the first-floor gallery; and one can well understand an unwillingness to mar the completeness of that wonderful room, with its collection of specimens probably unrivalled in the world.

The public certainly owes a debt of gratitude to the staff of the Natural History Museum for all they have done, in comparatively a few years' time, to popularise the collections. The use of a collection to the general public is at least doubled by the system of "introductory" specimens, and descriptive labels. With what advantage such labels might be placed on the cages at the Zoo, where one is left to wander round and learn what one can! One would almost be grateful at the Zoo for a return to the old system of the Wild Beast Show, where a keeper went round and lectured on the habits of the animals. Could not personally conducted parties be arranged by the authorities of the Natural History Museum? We have often noticed that the casual visitor is intensely grateful for any information afforded by students who may be working in the Museum.

CHAPTER II.

THE artificial reproduction of precious stones has nothing
in common with the *imitation* of gems. Ingenuity and skill,
and even a certain amount of scientific knowledge, have been
exercised in imitating diamonds, pearls, and so forth; but
the art is merely one of counterfeit, the materials produced
may deceive the eye, but they possess neither the chemical
composition nor the physical properties of the natural
objects which they simulate, except the qualities of colour
and lustre. In these two points the counterfeit is often
sufficiently good to deceive any save a practised eye; but
it must be borne in mind that the intrinsic value of a gem,
apart from the fictitious value due to rarity, depends not
solely on beauty of colour and lustre, but on the hardness
of the material which preserves, for instance, a cut ruby
from deterioration for centuries. It is the character of
permanence which gives to precious stones their pre-emi-
nent value among other beautiful objects.

The ruby may not only be imitated more or less successfully by colouring a dense and highly refracting kind of
glass; it can also be *reproduced*, that is to say, the thing
itself can be prepared in the laboratory.

Such a product is termed an *artificial* ruby, and the
common acceptation of the word appears to carry with it a
prejudice, as if it were intended to convey that the object
would be a ruby were it not that rubies are formed naturally,
whereas this was produced through the intervention and
contrivance of man. So much is this the case that if, as

may well happen, rubies should be produced of sufficient size for the purposes of the jeweller, there will certainly be a feeling against wearing the stones so produced as ornaments, due to the idea that the jewel is in some sort a sham. This is a natural but mistaken conception. A mineral—the ruby, for instance—is a body having a certain chemical composition and other important characteristics, such as those of its crystalline form, specific gravity and hardness. The body is formed through the operation of certain laws of chemical combination and of crystallisation. Whether the opportunity for the operation of these laws occurs in the bowels of the earth or in the crucible of the chemist, cannot be rightly held to affect the identity of the body produced.

The reproduction of minerals has been carried on for more than forty years, chiefly among a school of French chemists, and is a part of the great work of chemical synthesis.

Synthetical or constructive work only begins at an advanced stage in the study of an experimental science, and the synthesis of minerals was necessarily preceded by many years of analytical investigation.

In the first decade of the present century, the laws which regulate the proportions in which the elements enter into chemical combination were already established. At this time the art of chemical analysis was being rapidly developed, and its methods were applied to the examination of minerals. It was found that the elements they contained were present in those particular, definite, proportions which had been found to be characteristic of chemical combination. It was during this epoch also that the laws of crystallography were established.

The chemical composition and the crystalline form were recognised as the two most essential characteristics of each mineral species, although other properties were duly taken account of, as, for instance, specific gravity, hardness, and colour.

Thus mineralogy was put on a sound footing as a classificatory and analytical science; but the next step in advance, the introduction of synthesis, the building up of the minerals, appeared to be beyond the power of the experimentalist.

It was found that substances prepared in the laboratory, having the same chemical composition as natural minerals, did not possess their other characteristic features. Thus heavy spar has the same chemical composition as the sulphate of barium produced in the laboratory; but whereas the first is a hard, well-crystallised body, the second is formed as a fine powder, without coherence or crystalline form.

Again, silica occurs in nature as the well-known rock crystal, but in ordinary chemical processes it is obtained as a gritty powder.

The silicates, a class of substances comprising many well-crystallised gems, such as the garnet, could only be reproduced artificially as *glasses*, transparent indeed and coherent, but without crystalline structure.

Such failures gave rise to the impression that there was some special influence or force at work in Nature in the production of minerals which could not be commanded by the chemist, just as it was supposed that the substances produced in the vegetable and animal kingdoms needed the action of the so-called *vital force*, and were incapable of reproduction in the laboratory.

The belief in this *vital force* was dispelled when the advance of chemistry solved the problem of the synthesis of organic bodies. Similarly it was found that by modifying the ordinary methods of the chemical laboratory, so as to imitate more closely the conditions obtaining in the formation of rocks and of mineral veins, compounds could be produced, having not only the chemical composition, but the other characteristics of the natural minerals.

For instance, in the case of barium sulphate, the material is produced in the laboratory by the interaction of a solution of a barium salt and a solution of a sulphate. It was found that if special devices were adopted so that the two solutions only came in contact with extreme slowness, the forces of crystallisation came into play, and the barium sulphate separated out with the form, hardness, and other characteristics of the natural mineral. The processes previously employed had been too rough and hasty, and had not reproduced the conditions of Nature's laboratory.

Water plays a part in most of the ordinary processes of the chemical laboratory, but, under the usual conditions, water cannot be raised to a temperature above 100° C., since it is then converted into steam. In the depths of the earth, great pressure comes into action ; and when there is at the same time a high temperature, water, kept liquid by the pressure, acts under very special conditions. By heating silicates, such as glass, with water in strong steel vessels, so that a high temperature and great pressure are obtained, it is found that the silica is separated in the form of quartz, in crystals reproducing in the most complete manner the minute peculiarities, the surface-markings and striations, of the natural mineral.

For the production of corundum, a *flux* is employed, *i.e.* a substance which fuses at a moderate temperature, and in which the alumina dissolves, to separate out, on cooling, in the crystalline form. The colour of the ruby—one of the varieties of corundum—is due, not to the substance of which it is mainly composed, but to a very small proportion of a colouring matter. By the addition of a small amount of a suitable material, the red colour is obtained in the product of the laboratory, and, by varying the colouring material, sapphire and Oriental emerald have been obtained.

The size of the specimens obtained is small, one-third of a carat being about the maximum for rubies. The carat is equal to about $3\frac{1}{5}$ grains. A *cut* ruby weighing 1 grain would be suitable for one of the smaller stones of a ruby ring. In the process of cutting, however, the weight is generally reduced by one-half, so that the largest specimens yet produced are not adapted for employment as ornaments. They are, however, used in the jewelling of watches.

The details of the method, employed at the present time in the production of rubies, are as follows :—

Chemically precipitated amorphous alumina is heated with barium fluoride, or a mixture of the fluorides of the alkaline earths, which acts as a flux, and a trace of bichromate of potash is added to impart the red colour. The addition of carbonate of potash furthers the formation of larger crystals. The heating is kept up for several days, at the end of which time a plentiful crop of crystals is obtained. Although the aggregate weight obtained in one operation amounts to some pounds, the individual crystals are, as has been said, small in size.

It is frequently contended that the fact of reproduction is the only essential point, and that the size of the crystals produced is of little importance from the scientific point of view. It must, nevertheless, be allowed that the interest of this work will be much increased when products are obtained which will compare in size and beauty with those occurring in Nature.

Of other gems, some—as the garnet and the spinelles— have been prepared ; others, as the emerald, have hitherto proved less tractable.

In the case of turquoise, the artificially prepared substance has the chemical composition and the appearance of the natural stone; but inasmuch as the laboratory product behaves differently under certain conditions, as, for instance, when heated, it must be considered as an approximate reproduction only, if not looked upon as a mere imitation.

The pearl is formed of aragonite, a mineral readily re- produced by evaporating a hot solution of carbonate of lime. The peculiar beauty of the pearl is, however, due to the structure resulting from its mode of growth. It would be rash to hazard an opinion as to whether this structure could be imparted by methods at the disposal of the chemist.

But the great problem in the artificial production of gems

is the preparation of the diamond, and this problem is still unsolved. (See, however, later information on this subject in the essay on "The Making of the Diamond.")

Popular prejudice has relegated the attempt to produce the diamond to the same category as the endeavour of the alchemist to transmute the baser metals into gold.

The aim of the alchemist was once a legitimate object of scientific research. In the light of modern knowledge of the nature of chemical elements it is so no longer. The endeavour to obtain the element carbon in that transparent crystalline form in which it is found in Nature, has certainly nothing in common with the work of the alchemist. Yet the light in which the attempt is viewed by the majority is still (1891 A.D.) that so graphically described by Balzac in his ingenious novel, "La Recherche de l'Absolu."

Balthazar Claes devotes his life to the endeavour to re-produce the diamond, and "people would scarcely speak to him—a man in the nineteenth century seeking the philosopher's stone. They called him an alchemist, and said he might as well try to make gold. As he passed by in the street people pointed him out with expressions of pity or contempt."

The want of success which has hitherto attended the efforts of the Balthazars of real life is perhaps scarcely to be wondered at. In the case of other minerals the successful reproduction has generally been achieved only after the minute study of the mode of natural occurrence, which has afforded guidance as to the best means of imitating the natural process of formation. It is only of recent years that the diamond has been found in its original matrix, so that materials have hitherto been wanting on which to base experimental methods.

The chemical nature of the body, a combustible substance, is so different from that of the ruby and most other gems, which are oxides or oxidised materials, that the methods to be employed for its production will probably involve the application of different principles.

There is no reason, however, to regard the problem as insoluble. When sufficient guiding data have been obtained, skill will not be wanting to imitate in the laboratory the conditions under which Nature has worked in the formation of this most beautiful product of the mineral world.

The record of the investigation of a geological problem may generally be divided into two parts, the descriptive and the explanatory. A rock, for instance, is described according to its mode of occurrence, structure, and mineralogical composition; then follow deductions as to the epoch at which it was formed, and the mechanism of the actions by which its particular characters are supposed to have been produced. This, as a rule, marks the limit of the geologist's investigation of such a problem; far too seldom are the conclusions submitted to the test of experimental methods. This lack of the confirmatory evidence of experiment makes a large part of the literature of geology very unsatisfactory reading, the deductions being too often either indefinite or inconclusive. Mineral synthesis has done something to raise geology to the rank of an experimental science.

In the last years of the eighteenth century a controversy raged between the schools of Hutton and of Werner, as to whether heat or the action of water had been the dominating influence at work in the formation of the rocks of the earth's crust.

By what agency had chalk been converted into limestone or marble? How can this have been effected by heat, said the school of Werner, since heat decomposes carbonate of lime, expelling the carbonic acid?

The answer to this question was furnished by the experiments of Sir James Hall, "On the Action of Heat as modified by Pressure." Chalk was heated in a gun-barrel, the end of which was firmly closed. Under these conditions, the pressure increasing as the temperature is raised, the carbonic acid is not driven off from the carbonate of lime, the change induced being not chemical but physical, the powdery non-coherent chalk being converted into a compact crystalline mass, having all the characters of limestone, or of marble.

Hall also investigated another problem connected with the same controversy. Hutton maintained the purely igneous origin of those rocks which have characters similar to the modern lavas. It had, however, been noticed that if a piece of a crystalline rock were melted in a crucible, it was not reproduced on cooling, but that a uniform glassy mass was formed. By a judicious combination of the methods of observation and of experiment, Hall obtained important evidence as to the conditions of crystallisation of rocks. He observed during eruptions of lava that a great part of the crystallisation of the constituent minerals took place slowly during the gradual cooling of the mass of molten rock. Founding a method upon this observation, he melted various rocks in graphite crucibles, and maintained the materials in a state of fusion for a long time, taking care that the temperature should be somewhat above that necessary to melt the glassy mass. Crystals gradually formed, and a crystalline rock was reproduced, of which the melting point was higher than that of the glass formed in previous experiments where the cooling had been rapid.

Similar experiments were conducted about the same time (1804) by Gregory Watt. They were on a larger scale, a reverberatory furnace being employed in place of a crucible. The molten material was only allowed to cool with extreme slowness. From time to time samples were withdrawn and examined after solidification. Those in which the annealing process had continued longest were the most perfectly crystalline, and possessed the highest specific gravity, just as a natural crystalline rock, such as granite, is denser than a glassy rock (*e.g.* obsidian) of the same chemical composition.

These early experiments elucidated several important points with regard to the processes which have taken place in the formation of the eruptive rocks. The products obtained were, however, very imperfect reproductions of the natural rocks, and the methods for the determination of mineral species were at that time too rough to allow of the identification of the small and imperfect crystals obtained.

Before the date (1866) of the next important experimental research on the formation of rocks by igneous fusion, the application of the microscope in petrological work had effected a revolution in this respect. A slice of rock so thin as to be transparent, reveals to the microscope the outline, and even the internal structure, of the minute crystals which form its groundwork or *base*. The crystal of each mineral species shows its characteristics of form, the particular angles at which its faces are inclined to one another, and the lines developed in the process of grinding the thin section which indicate the directions of cleavage. Not less important in the work of identification are the optical characters which determine the tints assumed by different parts of the field of view according as the polarised light passes through the plate of one or other of the minerals of which the rock is composed. The refinements of optical analysis enable the identification of the species to be made with certainty even in crystals of microscopic size.

At the date to which we have referred, M. Daubrée published his experiments on the reproduction of the rocks of a certain class of meteorites. The meteorites being melted and kept for some time in the liquid condition, the constituent minerals began to crystallise out, and finally, after slow cooling, a rock was produced having the same constituent minerals as the original meteorite. Almost the only difference between the meteorites and the artificial products was the absence in the latter of that *brecciated* structure which frequently characterises an eruptive rock which has undergone violent mechanical strains.

By the employment of the modern refinements of microscopic and of chemical analysis, Daubrée was able to establish the absolute identity of the minerals contained in his artificial products with those of the meteorites. The class of meteorites for which the method of reproduction was found successful were those containing the smallest proportion of combined silica, characterised by the presence of olivine and augite and by the absence of the feldspars. Except for the presence of metallic iron, the mineral composition of these meteorites is very similar to that of what are termed the ultra-basic rocks, *i.e.* those the analysis of which shows the smallest proportion of silica. Many basalts and other lavas come under the category of ultra-basic or basic rocks.

Observational evidence appeared, however, to favour the view that these basalts and lavas had not been produced by the purely igneous method employed by Daubrée in the reproduction of meteorites, but that the action of water had played an important part in their formation.

It was in 1878 that MM. Fouqué and Lévy commenced the celebrated research in which they showed that the more basic eruptive rocks can be reproduced in every detail of mineral composition and structure by the action of heat alone without invoking the aid of pressure, or the intervention of water or of any other substance not forming a constituent of the rock.

To appreciate the details of their method, it is necessary to make clear the guiding data which were furnished by the study of the minute structure of rocks. Some eruptive rocks are entirely composed of an aggregate of perfectly crystallised minerals. One or more of the constituents (in granite, the quartz) may not show crystalline faces; they have presumably solidified last, and have been compelled to mould themselves round the crystals already formed, but their structure is completely crystalline, as is shown by their action on polarised light.

Other rocks differ from the *holocrystalline* in that the crystallised minerals are imbedded in a vitreous or glassy matrix which scarcely affects polarised light. These rocks are classed, from the character of the ground mass, as glassy rocks.

The most common structure of eruptive rocks is that of the third class, of which the ground mass has begun to crystallise before solidification, but the crystallisation has only gone as far as the production of *microliths*. These are crystals of small size, most frequently microscopic, which are so far developed that the determination of their species can be effected. They are seen to be grouped round the larger crystals of the rock in a manner plainly indicating their later formation. It appears reasonable to suppose that the microliths are formed during and after the welling up of the rock, whilst the formation of the large crystals may be referred to a previous epoch before the disturbance of the fluid mass from its subterranean position, when a condition of calm fusion favoured their growth and development. The temperature at which they were produced must be supposed to be higher than in the case of the microliths. Between these two epochs of crystallisation comes the eruption, during which the older crystals may be rounded, worn, or broken by shock.

Hall had shown that to obtain a crystalline structure instead of a glassy mass, it was necessary to keep the material at a temperature slightly above that of the melting point of the glass.

If, as appeared probable, the minerals of the different epochs of crystallisation did not possess the same degree or fusibility, it would be necessary in order to reproduce this association of minerals to maintain the materials at a series of temperatures successively decreasing. The result of the final operation might be expected to be the solidification of a mass of microliths of the more fusible minerals cementing together the larger crystals already formed. Such was the method employed by Fouqué and Lévy, and the result was in complete accordance with their expectations.

As an example of their work, we will describe the reproduction of a basalt precisely similar in character to certain basalts found in the Department of Auvergne. A mixture of substances prepared in the laboratory of the same

D

chemical composition as the rock was placed in a platinum crucible, which was maintained at a white heat for forty-eight hours. A sample taken out at the end of this time, and allowed to cool rapidly, showed, on solidifying, crystals of olivine imbedded in a brown-coloured, glassy matrix.

At the end of the first forty-eight hours the position of the crucible in the furnace was changed so that the temperature was lowered to that corresponding to a bright red heat, at which it was kept for a second period of forty-eight hours. The product obtained at the end of this time showed the crystals of olivine as before, but imbedded not in a glass but in a matrix composed of microliths of augite and of soda-lime feldspar.

Among the other rocks reproduced was a leucite-lava, the crystals of leucite having rounded angles just as in the natural rock showing that this peculiarity is not necessarily due, as has been supposed, to the effects of disturbance after the first epoch of crystallisation.

The rocks produced by the methods we have described are of the same character as those formed in the volcanic eruptions of the present time, which belong to the class of the more basic rocks.

The more siliceous rocks, such as granite, which contain free silica in the form of quartz, do not appear to be formed under the conditions obtaining in the eruptive processes which geologists have been able to observe in actual operation. When the materials of the *acid* rocks are subjected to the processes above described, the minerals which crystallise are not those of the original rock, but are of different crystalline form, even when they have the same chemical composition. The excess of silica remains in the uncombined state, but has characters resembling those of the variety known as tridymite rather than those of quartz. The acid rocks and their characteristic minerals (as quartz, potash-feldspar, and soda-feldspar) have probably been formed by processes radically different from those which produce the basic lavas.

These minerals have indeed been reproduced by the reaction of suitable materials in the presence of water, at a high temperature and pressure, but it has hitherto not been found possible to reproduce the compacted association of the minerals which constitutes an acid rock. Sufficient data have, however, been obtained to justify the belief that at no distant date the problem of the mode of formation of this class of rocks will be solved by the experimental method.

One of the most important contributions to experimental geology during recent years is the discovery of Spring that pressure is capable of inducing chemical change independently of its effect in raising the temperature of bodies. This discovery has a direct bearing upon the phenomena of *metamorphism*, or the bodily conversion of rocks into others of a completely different character.

Spring has shown that, by the application of great pressure, chemical combination is induced in cases where the compound occupies a smaller volume than the components, and conversely that a decomposition is brought about by pressure, when the volume of the bodies formed by decomposition is less than that of the compound. Briefly, pressure brings about such chemical changes as are accompanied by a contraction.

The apparatus employed in Spring's experiments consisted of a small steel chamber in which the substances were placed, furnished with a piston worked by a powerful lever, provided at the end with a heavy weight. If the piston were forced down rapidly the substances would be heated, and it would be impossible to discriminate between the changes due to rise in temperature and those due to increased pressure. In Spring's apparatus the lever is lowered very gradually, its descent being regulated by a finely cut screw. The steel chamber is surrounded by water, in which is placed a delicate thermometer, and the descent of the piston is operated so gradually that there is practically no rise of temperature. These experiments afford the first example of the direct conversion of mechanical work into chemical energy.

As a last example of the application of experimental methods in geology, we will deal with some of the problems presented by mineral veins.

The cracks and openings by which rocks are traversed are in some cases unfilled, in others they contain *débris* of the rock itself, and lastly, they are sometimes found filled with foreign minerals, and are then known as mineral veins. Most of the fine well-crystallised minerals which adorn museum collections come from mineral veins, or from cavities in rocks, known as "geodes." Tin veins, for instance, contain the oxide of tin, cassiterite, in large well-formed crystals. Others contain oxide of iron, also well crystallised; and another class contain the metallic sulphides, such as galena, the common lead ore found in Derbyshire and elsewhere.

The processes by which these minerals have accumulated in the veins, and the mode in which their crystallisation was induced, long remained a mystery.

In studying the characters of tin veins, Daubrée was struck by the constant presence of minerals, such as apatite, topaz, and tourmaline, which contain the elements chlorine and fluorine. It is known that the chloride and fluoride of tin are volatile, and that these compounds are decomposed by water, the hydrogen of the water forming hydrochloric or hydrofluoric acid gas and the oxygen combining with the metal.

Daubrée's experiments showed that if the vapour of water be brought in contact with that of chloride of tin at a fairly high temperature the oxide of the metal is formed in crystals, having all the characters of the mineral cassierite.

Subsequently Sainte-Claire Deville showed that by passing hydrochloric acid gas over strongly heated oxide of tin, the chloride of the metal was formed, and steam. These two gases are carried on by the current of hydrochloric acid gas, and in a somewhat cooler part of the apparatus react again upon one another, re-forming hydrochloric acid, and depositing the oxide of tin in the crystalline form.

Hydrofluoric acid and other re-agents which have been found to act in a similar manner have been termed *mineralisateurs.* They are capable of effecting the transport of non-volatile substances, such as oxide of tin, depositing them in the crystalline form, and, their work done, they leave no trace of their former presence except in the combination of their more active elements with constituents of the surrounding rock or of certain minerals of the vein, the *gangue* minerals, as they are termed.

One of the commonest of the *gangue* minerals is calcite, and by passing the vapour of chloride of phosphorus over heated calcite, Daubrée found that apatite is formed, a mineral which, as has been mentioned, is characteristic of tin veins.

In the formation of the metallic sulphides, sulphuretted hydrogen has played the part of *mineralisateur.*

In some cases, no doubt, the mineralising action is effected by the vapour of the active substance ascending from the heated interior; in other cases, as in the neighbourhood of the great Compstock Lode, the substances are in solution, subterranean streams saturated with sulphuretted hydrogen and other powerful solvents serving to extract the metallic compounds from the rocks in which they are present in small quantities and concentrating them in the veins. By such processes metallic ores are collected in the rock veins, which thus become the great store-houses of mineral wealth.

CHAPTER III.

DIAMOND MINING AND DIAMOND MAKING.

SECT. I.—THE DIAMOND MINES OF SOUTH AFRICA.

THE first South African diamond was found in 1867, and during the next three years diamonds were obtained from the river workings.

In the South African mines the diamond appears to occur in its proper matrix. In 1870, the *mother rock* was found at Kimberley. This rock occurs in pipes, as they are termed, round or oval funnels with a surface area of several acres, and of great but unknown depth. Some have already been excavated to a depth of eight hundred feet or more without any sign of reaching a bottom, or bed rock. The rock at and near the surface is termed the yellow ground, a friable material from which the diamonds are readily extracted. When the yellow ground was worked through and the *blue rock* struck, many of the miners imagined that the deposit was worked out, and abandoned their claims. Others filled in the excavations with the *yellow*, and sold the claims to less knowing but more fortunate adventurers.

The "blue" proved to be the real matrix of the diamond, the "yellow" being merely the blue rock altered by weathering, a yellow colour having been produced by formation of oxide of iron from the action of the atmosphere on the highly ferruginous materials.

The "blue" is a volcanic rock of very peculiar character, extremely heavy, and of the structure known as *brecciated*, characteristic of a volcanic rock which has been subjected to movement after hardening.   It contains boulders of all sizes up to twenty tons, and pieces of sandstone, shale, and occasionally fragments of fossil trees.   A detailed study leaves no doubt that the *pipes* are of eruptive origin—a pipe being, in fact, the neck of an old volcano—and that the blue was forced up from below, the sandstone, and so forth, which show evidence of aqueous action, being derived from the material of the formerly existing rock.

The extraction of diamonds from the blue is less readily effected than from the yellow surface material.   The blue rock is first spread out on the surface and exposed to the action of the weather.   The disintegration proceeds best when wet and fine days alternate, and in dry weather the process is hastened by watering the material.   The change which occurs is the same as that which produced the yellow at the surface of the pipes.   When the process has gone on sufficiently long, the rock is treated by washing machinery.   The lighter materials are washed away in a pulsator, and the diamonds and heavier minerals, such as pyrites and garnet, are left in a sort of mud, which is brought on to a table and carefully searched.

Comparatively few diamonds are discovered during the actual mining of the rock ; they are imbedded singly, and are not conspicuous objects.   The general appearance of the natural diamond is somewhat like that of a piece of white gum—the brilliancy of the stone being only called out by the operation of cutting.   Naturally the stones most likely to be noticed are those of large size, and these are the great prizes of diamond mining, since the value of a stone rises in a very rapid ratio with the size.

Now that the mining is no longer done by private adventurers, but the whole worked under the management of companies, it is extremely important to ensure that the workers shall not secrete the stones.   The miners, when

they come up at the end of their shift, are carefully searched.
As the natives work without clothes, it might be supposed
that the searching would be a very simple matter; it is not
so, however, the hair, ears, nose, and teeth being used as
places of concealment. Moreover, the stones were frequently swallowed, which led to the system of keeping the
native workers in compounds, which they are not allowed
to leave during the term of their engagement. Most of
the work at the larger mines is done by contract, the company paying so much a load, and providing and housing
the native labourers, who are paid by the person under-
taking the contract—generally himself an experienced
miner.

Now that the mining is carried on at great depths, the
risks of accident are considerable. Till a depth of about four
hundred feet was reached the system of open workings was
followed, but as the depth increased the falls of the blue
rock became more frequent. Zulu watchers were stationed
to give warning to the miners when the rock showed signs
of giving way. This kind of work, involving constant
alertness during long periods of inaction, is peculiarly
trying to white men, but is admirably performed by the
Zulus.

The appearance of the mines from above is that of huge
craters, at the bottom of which the tunnelling and shaft-
sinking commences. The mode of descent into these huge
craters is simple and expeditious, but not suited to nervous
passengers. A kind of truck or tub is suspended below
two wire ropes, the flanged wheels, which are, of course,
above the tub, running on the ropes. The tub is attached
by a third wire rope to a winding engine, the rate of
winding being about forty feet a second, or nearly thirty
miles an hour.

The average weight of diamonds obtained per load (16
cubic feet) of blue rock varies greatly in the different
mines; the ordinary limits may be put at $\frac{1}{2}$ and $2\frac{1}{2}$ carats,
Those mines which have the largest number of stones do

not generally produce the best quality, so that in the matter of profit there is a sort of compensating arrangement. The finest diamonds at present in the market come from South Africa—notwithstanding the popular prejudice which till quite recently assigned all diamonds of "the first water" (a term, by the way, which is not customary among diamond merchants, though dear to novelists) to the mines of Golconda or Brazil. Diamonds, exceeding the Koh-i-noor in size and equal in brilliancy, have been found in the South African mines, but such stones are no longer sought after. Their price, calculated to rise rather more rapidly than the square of the weight, is nominally very great, but no one will pay the price ; and, strange as it may seem, such stones are now split up into two or three of the largest size that are ordinarily worn. The "crowned heads" are apparently now all furnished with crowns, and diamonds of a size which seems only suitable for regalia can no longer be readily disposed of.

To the mineralogist the chief interest of the South African mines lies in the fact that the blue rock, or "kimberlite," appears to be the original matrix of the diamond. Till of late years the diamond had only been found in alluvial deposits, a mode of occurrence which gives but little indication of the manner of its formation. In kimberlite, however, it appears *in situ,* and the character of the minerals with which it is associated may perhaps afford some guidance as to the means to be adopted for the reproduction of the diamond. The rock belongs to the class termed ultra-basic, having a low percentage of silica and a high specific gravity. The rock is analogous to certain meteorites.

The following are the principal minerals of the rock :—

Biotite	Ilmenite
Bronzite	Olivine
Chrome diallage	Perowskite
Chrome iron ore	Pyrites
Garnet	Smaragdite
Graphite	

The olivine occurs in large quantity. This mineral, under the action of weathering, is decomposed, forming serpentine.

In studying the occurrence of diamond-bearing sand and deposits in different parts of the world, the late Professor Carvel Lewis arrived at the conclusion that diamond-bearing deposits occur, as a rule, in watercourses which take their origin in mountainous tracts characterised by the presence of serpentine. Serpentine, or a rock which weathers to serpentine, was considered by Lewis to be the real matrix of the diamond. The position and mineralogical character of the kimberlite rock, filling in the neck of a volcanic vent, plainly show its igneous origin, and the fact that it has been protruded from below. It is not definitely known whether the diamonds were already formed in the rock before its eruption, or whether they have been produced by alteration of the materials contained in the rock displaced by the eruption.

It is worthy of notice, however, that a black shale forms one of the surrounding rocks, and pieces of this shale have been found baked and otherwise altered in the " blue rock." The suggestion has been thrown out that the diamonds have been formed by the alteration of the carbonaceous matter of the shale under the influence of a moderately high temperature and great pressure.

Such indications of origin are useful as affording suggestions to the experimentalist, to whom, in spite of previous failures, we must still look to tell us definitely how the diamond is formed.

SECT. II.—THE MAKING OF THE DIAMOND.

(Written in March, 1894.)

The reproduction of the diamond by M. Moissan has put the coping-stone to the work of mineralogical synthesis. For some years past it has been thought that the solution of this problem was merely a matter of time and patience ; but it is no little satisfaction to be able to say at last that the thing has been done, for it is indeed a striking illustration of the power over stubborn matter which has been won by the students of science.   In the light of what has now been accomplished, it may not be without interest to refer to what was written by the present author on the subject of the production of diamonds previously to the publication of M. Moissan's work.   In May, 1891, in the conclusion of an essay on "Mineral Synthesis," the matter was referred to as follows :—

"The great problem in the artificial production of gems is the preparation of the diamond. . . . In the case of other minerals the successful production has generally only been achieved after a minute study of the mode of natural occurrence, and this has afforded guidance as to the best means of imitating the natural process of formation.   It is only of recent years that the diamond has been found in its original matrix, so that materials have been wanting on which to base experimental methods.   The chemical nature of the body, a combustible substance, is so different from that of the ruby and most other gems, which are oxides or oxidised materials, that the methods to be employed for its production will probably involve the application of different principles.   There is no reason, however, to regard the problem as insoluble.   When sufficient guiding data have been obtained, skill will not be wanting to imitate in the laboratory the conditions under which Nature has worked in the formation of this most beautiful product of the mineral world."

What some of these determining conditions might be was indicated in a subsequent paper on " The Diamond Mines of South Africa," which first appeared in October, 1891.   " To the mineralogist the chief interest of the South African mines lies in the fact that the 'blue rock' or kimberlite appears to be the original matrix of the diamond. . . . It is worthy of note that a black shale forms one of the surrounding rocks, and pieces of this shale have been found baked and otherwise altered in the blue rock.   The suggestion has been thrown out that the diamonds were formed by the alteration of the carbonaceous matter of the shale under the influence of a moderately high temperature and great pressure. Such indications are useful as affording suggestions to the experimentalist, to whom, in spite of previous failures, we must look to tell us definitely how the diamond is formed."

If the diamond be highly heated in the presence of oxygen it takes fire, as is well known, and burns with the formation of carbonic acid.

If it be heated not in contact with oxygen it swells up and blackens, reverting to the ordinary charred form of carbon.

But the action of heat upon bodies is in many cases very different when they are subjected to high pressure—a principle established by Sir John Hall more than one hundred years ago in his celebrated research on the conversion of chalk into marble.

As will be seen, M. Moissan invoked the aid of pressure to modify the action of heat in his experiments, and produced diamonds from charcoal, a substance of the same nature as the "shale" which occurs in the Kimberley rock.

The formation of crystals is, as a rule, best brought about either by sublimation or by cooling a solution. Carbon, however, cannot be distilled or sublimed, and is insoluble in all ordinary solvents, such as water or aqueous solutions of acids and alkalies, or such as alcohol, ether or benzene. On the other hand, molten metals can take up or dissolve carbon to a not inconsiderable extent, as happens, for instance, in the process of iron-smelting. The molten iron in the blast furnace dissolves some of the carbonaceous fuel, a part of which, when the iron is allowed to cool and solidify, crystallises out in plates of graphite.

This is an example of the production of a crystalline form of carbon from a non-crystalline variety, and it is at the same time an instance of the artificial formation of a mineral.

M. Moissan, in his experiments, employed iron as a solvent for carbon, which was used in the form of charcoal; but he modified the action of heat and of the solvent by subjecting the carbon-saturated iron to considerable pressure.

It may be noted here that M. Moissan finds the principal constituent in the ash of the native diamond to be oxide of iron. It is known also that native diamonds often contain liquefied gases in cavities of the crystal, and that they are sometimes liable to spontaneous disruption, owing to a state of strain which is probably due to their having been formed under high pressure.

Iron melted by means of an electric furnace, and raised to a white heat, was allowed to saturate itself with carbon in the form of strongly compressed sugar charcoal. The crucible in which the operation was conducted was then plunged into cold water, which cools the outer part of the metal so as to form an outer layer of solid iron. While this outer coating is still red hot the crucible is withdrawn from the water, and the cooling proceeds more slowly.

To realize what goes on within the jacket of solid iron, we must remember that the still liquid interior is molten iron, containing a large excess of dissolved carbon, and that iron *expands* in the process of solidifying. Hence, during the process of solidification within the jacket or crust of chilled metal, great pressure is exerted. The process of solidification, therefore, goes on slowly and under great pressure, and examination of the resulting product showed that, under these changed conditions a part only of the surplus carbon had crystallised out as graphite, and that in the residue left after dissolving away all the iron by means of boiling hydrochloric acid and other solvents there was a certain quantity of a denser form of carbon (having a specific gravity of 3 to 3·5), and hard enough to scratch a ruby; and that among these heavier portions of the residue were transparent particles, having a greasy or waxy lustre, and marked with parallel striæ and triangular depressions. These transparent particles burnt when heated to 1050° C. in oxygen gas, and as it appeared, with the formation of carbonic acid; but the particles were too small to allow of a quantitative experiment. Similar results were obtained by the slightly modified method of rapidly cooling an ingot of molten iron saturated with carbon from a temperature of 2000° C. In a few cases small fragments were obtained "*qu'ils ressemblent aux petites fragments de diamant transparents que nous avons rencontrés dans la 'terre bleu' du Cap*" (*Comptes Rendus*, February 6, 1894).

The result may be summed up by saying that, up to the date of the experiments described in the above quoted paper, M. Moissan appears to have succeeded in *reproducing that transparent variety of carbon of which native diamonds are composed.* The specimens could hardly be called *diamonds*, although they showed certain characters of the native diamonds—*e.g.* a waxy lustre, and parallel striæ and triangular depressions on the surface.

Since the experiments above described, a happy modification of the method has given results of a far superior kind, tolerably perfect diamonds being formed, having the distinctive physical peculiarities of the native stone, and of sufficient size for M. Moissan to prove by quantitative chemical experiments upon some of the specimens that they burnt with the formation of pure carbonic acid. In the course of experiments made in former years by other experimenters using other methods, transparent crystalline bodies were obtained which were thought to be diamonds, until their failure to satisfy the carbonic acid test showed that the crystalline particles were not composed of carbon.

Moissan's modified method is as follows: Iron is saturated with carbon at the white heat of an electric furnace, and under pressure. The crucible containing the molten iron is then quickly lowered to the bottom of a bath of melted lead. This ensures quicker cooling than when the iron is plunged in water, owing to the tact, first, that the white-hot iron does not really come into contact with the water, and secondly, that the lead is a good conductor and carries away the heat rapidly. It seems that the two liquid metals behave towards one another much as oil and hot water, the molten iron collects in spherical globules which rise to the surface of the molten lead, the difference in the specific gravity of molten iron and of molten lead being considerable.

The surface of the drops of liquid iron which float upon the lead quickly solidifies, the smaller drops with a diameter of one to two centimètres first, the larger drops after a lapse of a longer time; and the little solid balls of iron are left to float on the molten lead, where they cool down.

The interior of the balls is of course liquid long after the formation of the solid crust. The tendency of the central parts to solidify is resisted by the solid crust, owing to the fact before mentioned, that iron expands in the act of solidification. Meanwhile a part of the carbon crystallizes out from its solution in the liquid iron. After a time, as the cooling goes on, the lead also solidifies, and the little iron balls are left imbedded in the ingot of lead. Then begins the process of getting at the small quantity of the carbonaceous material which it is desired to examine. The lead which adheres to the iron is dissolved away by nitric acid, the iron itself is dissolved by hydrochloric acid, and further treatment with suitable solvents leaves the sought-for residue, a small quantity of material left after the tedious process of removing by slow chemical means the relatively large mass of metal.

Transparent diamonds are found in the residue, having well-defined crystalline faces, striated and marked in the well-known way, and the edges generally curved; they have the high refracting power, the specific gravity, and the hardness of the native stone. The peculiar form known as the *hemihedral* predominates amongst these crystals as in those of native diamonds, and their formation under pressure is found to give rise to the phenomena of anomalous polarisation of the light which passes through them, as well as occasionally to spontaneous disruption; characters which are sometimes noticed in the native stone.

The diamonds are, of course, small ; one with a diameter of half a millimètre appears to be reckoned a fine specimen. Further practice in working the process will probably enable larger specimens to be produced, as has been the case with the production of rubies.

However this may be, the production of diamond is now an accomplished fact, achieved by the patient skill of the same worker who, seven years before, successfully overcame the great experimental difficulties which had rendered fruitless the many former attempts to isolate the element fluorine.

CHEMISTRY.

CHAPTER IV.

ELEMENTS AND ATOMS.

SECT. I.—LAVOISIER AND DALTON.

THE older conception of a chemical element was that of a property, or group of properties, which, being common to several substances, was regarded as a *principle* existing in all those substances. Whereas, we now think of the elements as the undecomposed residues of natural substances, and as forms of ponderable matter uncreatable and indestructible, the earlier conception of a chemical principle was that of a constant *property* possessed by certain forms of matter, *e.g.* the combustibility which is a property common to substances of animal and vegetable growth.

It is the dependence of certain properties, on the presence of certain kinds of substance, which aids us in many cases to connect the older chemistry with that of the present time, by identifying some of the so-called *principles* of the seventeenth century with the corresponding *elements* of the nineteenth-century chemists.

A short history of the study of carbon will serve to show the gradual evolution of modern ideas about chemical elements.

When vegetable or animal materials are heated with a limited access of air, the result is that they are charred. Materials differing in almost every respect, except in their being formed in the processes of animal or plant life, agree in this—that when thus treated, they yield a *char*.

The word *carbo*, formed from the same root, appears to have been used in the Augustan age in the same sense, namely, to designate the char left by the partial burning of animal and vegetable bodies. It was usually applied to the char obtained from wood, or wood charcoal. In the more modern use of the Latin, *carbo* generally means coal, the vegetable origin of which is readily recognised.

The first important generalisation in the history of chemistry is that contained in the theory of phlogiston, which retained its hold upon the minds of chemists till the later years of the eighteenth century. The merit of this system lay in the fact that the various phenomena of oxidation were for the first time grouped together and referred to the same agency.

The change of properties undergone by metals on calcination was a favourite subject of study with the phlogistic chemists. An equal share of their attention was devoted to the means of restoring to calxes (or oxides) the metallic properties. The phlogistic chemists recognised that the process of calcination of metals was essentially a process of burning. The char, or substance left by the partial burning of animal and vegetable bodies, constitutes the most generally applicable material for the preparation of the metals from their ores, or from the calxes. The calx or ore is mixed with charcoal, or with the analogous material coal, and these being heated together, the charcoal or coal is consumed, to all appearance ceases to exist, and a metal is produced from the calx.

It is to be observed that these changes only take place at a high temperature, and that the materials must therefore be heated. Now the primitive method of heating is to kindle a mass of vegetable matter, such as wood, burning the material with a free supply of air. It was, however, found that for metallurgical operations it was more advantageous to employ the prepared material charcoal, instead of the crude material wood. In metallurgical operations the charcoal (or the analogous material coal) plays two parts, each of which is essential to the reduction of the ore or calx. The first of these functions is the production of a high temperature; the second that of acting on the heated ore so as to form a metallic substance. The charcoal employed as fuel has undergone combustion, and the charcoal mixed with the ore has apparently imparted to the ore the power of burning, that is to say, has formed from it a metal, which is a material capable of burning. Coal, coke, wood charcoal, animal charcoal, lampblack — in short, all the various forms of char—appeared to possess in an eminent degree what was termed the "principle of combustibility." The word *carbone* was employed to denote the combustible *principle* contained in the various forms of char. The principle thus named is represented in terms of our modern views by what is known as a chemical element, and to this element the name *carbon* is given.

The phlogistic chemists made no real advance towards explaining the reducing action of the various bodies containing the principle or substance carbon. This explanation was reserved for Lavoisier, the great opponent of the phlogistic system. Lavoisier recognised the all-important fact that chemical transformation is not accompanied by any change of mass, and by weighing the several substances taking part in, or formed during, a chemical change, he was able to trace a constituent through a series of transformations.

Previous to the work of Lavoisier, Black had prepared and studied the gas called "fixed air" (carbonic acid), which is obtained by the action of an acid on limestone. Lavoisier found that the same gas is formed when charcoal is burnt, and he further proved that in this process a constituent of the air (oxygen) unites with the charcoal, the fixed air or carbonic acid formed being composed of *carbon* and oxygen.

Lavoisier likewise proved that in calcination the substance he termed oxygen is abstracted from the air, and combines with the metal, forming a calx or oxide. Finally, he showed that if a calx be heated with charcoal in the absence of air, carbonic acid is formed, proving that the change which has taken place consists in removing the substance oxygen from its combination with the metal.

From this series of experiments it was evident that the *reducing* function of any form of char is better expressed by saying that the substance carbon removes the substance oxygen than by saying that the combustible principle carbon imparts to a calx the principle of combustibility. These two modes of expressing the same facts may be represented thus—

Calx *minus* the substance oxygen = metal, and
Calx *plus* the principle of combustibility = metal.

From the date of Lavoisier's discovery, the formation of carbonic acid has been invariably employed for detecting the presence of the element carbon. Carbonic acid gas is readily recognised, even in the smallest traces, by the well-known property of throwing down a sediment of chalk when passed into lime water. Any substance which on burning forms carbonic acid contains the element carbon. The amount of carbonic acid formed, which can be accurately determined, supplies the best means of estimating the weight of carbon contained in any material.

If a material, when completely burnt, yields carbonic acid, *and nothing else,* then we say that the material is composed wholly of the element carbon. The purest wood charcoal, when completely burnt, yields no gas except carbonic acid, and no residue except a very small quantity of ash, in which can be recognised those mineral ingredients which the plant, from which the charcoal was prepared, derived from the soil. Wood charcoal consists, therefore, almost wholly of carbon. Any other form of char, when freed as far as possible from foreign matter (*e.g.* by drying to remove moisture), possesses the same, or nearly the same, physical constants as wood charcoal. The specific gravity is about 1·9. The materials are amorphous, *i.e.* without crystalline form.

It was not long after his discovery of the composition of carbonic acid that Lavoisier, in conjunction with other chemists, experimented upon the combustion of the diamond. The diamond was placed in a glass vessel containing air, and, by means of a powerful lens, the rays of the sun were concentrated to a focus on the diamond. The diamond was by this means heated sufficiently to burn in the air of the vessel, and the gas evolved was collected over mercury, and tested. It was found to be carbonic acid. Soon afterwards (1796) an English chemist showed that the amount of carbonic acid formed by burning equal weights of diamond and of pure charcoal is the same—a conclusion which has since been repeatedly verified with the superior accuracy of modern methods.

Both diamond and pure charcoal, therefore, consist wholly of the element carbon.

In the year 1800, a third substance was added to the list of substances known as forms of carbon. The mineral plumbago, or graphite, was formerly regarded as identical with molybdenum (a metallic sulphide), the appearance of the two substances being similar, and both possessing the property of marking paper with a black streak, whence the name graphite (γράφω, to write). It was shown, however, by Mackenzie, that graphite burns with formation of carbonic acid, the amount formed from a given weight of the material being the same as in the case of charcoal and of diamond. Here, then, we have a third *form of carbon*.

These three substances differ, in the first place, in certain important physical characters. The specific gravities are different, diamond standing highest in the list, and charcoal lowest. Diamond crystallises in the regular or cubic system, graphite in the hexagonal system, whilst charcoal has no crystalline form or structure.

But it is not only in physical properties that the three

substances differ ; they differ to a certain extent also in chemical character. The temperature at which diamond burns is much higher than that at which the combustion of charcoal takes place. Thus, although oxygen unites with both substances, and in each case forms the same product, yet the readiness with which this combination takes place is very different in the two cases. In other words, the chemical relations of diamond and of charcoal, with respect to oxygen, are by no means identical. Again, graphite differs from either of the foregoing, in that the combined action of nitric acid and potassium chlorate converts it into a peculiar acid, a solid substance, known as graphitic acid.

In spite of these marked differences, chemical as well as physical, it is the universal practice to denominate all three substances, diamond, graphite, and charcoal, as " forms of the element carbon," or *allotropic* modifications of carbon. It must be confessed that the present phraseology is not as clear as might be desired, and is constantly a stumbling-block in the way of the tyro in chemistry.

Set a schoolboy to write an essay on " allotropy" (or the existence of elementary substances in different " forms "), and he will choose as his example the element carbon. He will begin by pointing out how widely different are the substances, diamond, graphite, and charcoal, and will wind up his essay by saying that, notwithstanding these differences, " they are really the same thing—carbon."

The source of confusion must be explained by reference to the atomic theory. We possess a great mass of evidence to show that what we observe in any chemical process is the sum total as observed on the large scale of a number of phenomena, all precisely alike, occurring between the ultimate particles or chemical atoms of substances. All the thousands of known substances are formed by various combinations of atoms of a comparatively small number of chemical elements. All the atoms of any one chemical element are exactly alike, but are different (*e.g.*

in their mass) from the atom of any other element. A substance containing more than one kind of chemical atom is termed a compound substance or chemical compound. A substance containing only one kind of chemical atom is termed an elementary substance. Such a substance is diamond. It is formed wholly from atoms of one kind; carbon atoms. Charcoal likewise is formed wholly of carbon atoms. All carbon atoms are, we believe, alike; but this by no means necessitates the identity of substances composed wholly of those atoms. We must look upon diamond and charcoal as *structures* both formed of the same material (the carbon atom), but built up in different ways.

Regarded in this way, the subject of *allotropy* is easy to understand; as easy as it is to understand, for instance, that a certain Jacobean house in the north of England is not a Norman castle, though built of the sandstone blocks which once formed the feudal fortress originally standing on the same site. The same sandstone blocks served to build both the house and the castle, and in the same way diamond, charcoal, and graphite are constructed wholly of carbon atoms; but to say simply and without qualification that diamond, charcoal, and graphite are the same substance is a misleading expression.

It is to John Dalton that we owe the conception of the chemical atom. Sir Henry Roscoe's life of John Dalton explains the essential differences between Dalton's theory of chemical atoms and the older philosophic notions of a corpuscular or atomic constitution of matter. How far Dalton substantiated by his experiments the admirable theory which he formulated is a point less easy to decide; but it is upon this that his claim to rank with the greatest men of science must largely depend. Sir Henry Roscoe is an ardent Daltonist. "Dalton," he says, "is the founder of modern chemistry." In France Lavoisier is generally considered to hold this position. "La chimie," says a French writer, "est une science Française." Some English chemists also would be disposed to traverse the statement that Dalton was "the first to introduce the idea of quantity into chemistry." However this may be, the author of the hypothesis of chemical atoms has ample claims to the gratitude of the scientific world, and his brilliant theory has now been abundantly confirmed by the work of three generations of chemists.

Newton, and many philosophers before his time, had held that matter is composed of indivisible atoms. Other philosophers had favoured the doctrine of continuous, rather than corpuscular or atomic, matter. Dalton's mind was of the "corpuscular" turn. His opinions gained in clearness and precision by long and careful study of the physical properties of gases. He experimented upon the phenomena of gaseous diffusion, and found that a heavier and a lighter gas will intermix in opposition to the force of gravity. This, and other phenomena—*e.g.* of the absorption of gases by water, and of evaporation—pointed clearly to an atomistic constitution of gases. Dr. Henry, of Manchester, says, "It was in contemplating the essential conditions of elastic fluidity that Dalton first distinctly pictured to himself the existence of atoms." It had been shown by other chemists that the chemical combination of any two bodies takes place in a fixed proportion by weight, and that the proportions in which two bodies combine are equivalent to one another in all the chemical reactions in which those bodies are capable of taking part. Tables of the equivalent weights of chemical substances had been drawn up by Cavendish and others. It is Dalton's chief claim to admiration that he grasped the fact that the "equivalent weights" of the chemist are proportional to the weights of the philosopher's "atoms." That the atoms of different substances are endowed with different, and specific, weights, and that the relative weights of the atoms can be experimentally determined, was, we believe, a conception quite new to science. The theory of chemical atoms stands on a higher plane than all previous speculations as to the atomistic constitution of matter. As we have said, chemical combination, as observed in the laboratory, is, in Dalton's view, the sum total of a vast number of combinations, all precisely alike, occurring between the ultimate particles, or atoms, of the substances. Dalton's announcement of the law of *multiple proportions* helped to carry conviction of the truth of his atomic theory, and gave force and precision to his ideas. He found, or thought he found, that when an element combines chemically in more than one proportion with a second element, the combining weight of the second element is *twice* as great in one compound as in the other. This naturally follows if the atoms have specific weights, and if the simplest combination of the elements be atom with atom, and the next combination be of one atom with two atoms. The experimental numbers on which Dalton relied in support of his "law" of multiple proportions were, however, not such as to carry conviction except to a mind already biased. Both the "law" of multiple proportions and the atomic theory were enunciated by Dalton, but their truth was, in our opinion, established by other and more exact workers.* Dalton was a bold guesser, but an inexact experimentalist. His standard of accuracy in measurement was below the average of his own day, and far below that of Cavendish. It is hard to know which most to admire, Dalton's genius or his good fortune.

A New View of the Origin of Dalton's Atomic Theory.

The recent discovery of John Dalton's note-books has enabled Sir Henry Roscoe and Doctor Harden, in their "New View of Dalton's Atomic Theory," to trace the mental process by which the theory of chemical atoms was evolved. There is always a philosophic interest in tracing the steps by which a great thinker has been led

* The experimental basis of Dalton's Atomic Theory is treated more fully in the present writer's "Practical Proofs of Chemical Laws."

to formulate an important theory, and in the present case the interest is increased by the fact that there has long been current a circumstantial, but, as now appears, inaccurate, account of the origin of the atomic theory. Dalton himself did not publish any statement as to the steps by which he arrived at his great generalisation, and chemists have generally, although not universally, accepted the version given by Thomas Thomson, in his well-known "History of Chemistry." Dalton published the first provisional list of "atomic weights" in September, 1803. In August, 1804, Thomson spent a day or two in his company, and recorded the result of the visit in his book, which was published three years later. "Mr. Dalton," he says, "informed me that the atomic theory first occurred to him during his investigations of olefiant gas and carburetted hydrogen gas, at that time imperfectly understood, and the constitution of which was first fully developed by Mr. Dalton himself. It was obvious, from the experiments which he made upon them, that the constituents of both were carbon and hydrogen, and nothing else. He found, further, that if we reckon the carbon in each the same, then carburetted hydrogen contains exactly twice as much hydrogen as olefiant gas does. This determined him to state the ratios of these constituents in numbers, and to consider the olefiant gas a compound of one atom of carbon and one atom of hydrogen; and carburetted hydrogen of one atom of carbon and two atoms of hydrogen."

Thomas Thomson was a very inaccurate man in his experimental work, but unfortunately wrote in a pleasant and interesting manner, and his "History of Chemistry" has been, and still is, very widely read. It now appears as if his trustworthiness as an historian was on a par with his accuracy as an experimentalist. Already Dr. Debus, in a work published in 1894, had shown cause for doubting Thomson's version of the matter; and the "Life of Dalton," by his contemporary, Dr. Henry, gives an account somewhat different from Thomson's. The discovery of Dalton's

MS. confirms the statements of Henry and some of the conclusions of Debus, and goes much beyond them in the completeness with which it enables us to reconstruct the early history of the atomic theory. Briefly, it is found that the atomic theory did *not* originate from the discovery of an example of the law of combination in multiple proportions (as stated by Thomson), but contrariwise. Dalton approached the chemical question from the physical side. He was at the outset a physicist rather than a chemist. Following Newton, he supposed gases to be composed of numerous indivisible particles, and he developed this idea during his work upon mixed gases by supposing the particle or atom of each individual gas to have its own proper size and weight. This idea he *afterwards* applied in his consideration of chemical combination, which he supposed to take place by the joining up of atoms. What is observed in a chemical process is, on Dalton's supposition, the total of a vast number of occurrences, all exactly alike, which take place between the atoms. The proportions by weight in which elements combine chemically he considered to be the relative weights of their atoms. In those cases where two substances were known to combine in more than one proportion, he was, in accordance with his theory, obliged to represent the quantity of one element in the second compound as a whole multiple of that in the first. The theory accorded perfectly with the already known law of the constant composition of chemical compounds, and of those relations between the composition of different compounds which were expressed in the tables of "equivalent weights." The fact established by Proust that in a series of compounds containing the same two elements the proportions do not change gradually, but *per saltum*, by leaps and bounds, was also in accordance with Dalton's view. Here, however, the theory was in advance of experiment, for theory indicated that the proportions should be numerically simple ; "multiple proportion" was a necessary consequence of the theory. Existing analyses, examined by the light of the theory, were found to yield a certain amount of confirmation of this theoretical deduction; and when more experimental evidence had been gained, the hypothesis of combination in multiple proportions was raised to the rank of an experimental law. It is the theory of atoms which pointed the way to the discovery of the law, and not, as has been generally thought, the discovery of the law which suggested the theory. The successful explanation of the increase of weight *per saltum* in a series of compounds of the same element was one of the earliest triumphs of the new theory.

SECT. II.—STAS'S RESEARCHES UPON COMBINING PROPORTIONS.

The name of Stas has been a household word among chemists for nearly half a century, and his writings, the celebrated "Recherches sur les Lois des Proportions Chimiques," are among the canonical books of chemistry. In all that relates to the experimental art Stas stands unsurpassed. The marvellous patience with which he matured his methods, and the skilful care with which the final experiments were carried out, stand recorded in his classical memoirs with that clearness and precision of expression characteristic of scientific writings in the French language.

Stas's work bore on one subject only—the determination of the relative weights of the atoms, with a view more particularly to ascertain if there existed any simple relation between the weights of the chemical atoms.

In order to explain how this investigation came to be the mission of Stas's life, we must refer to the state of chemical theory in the second decade of the present century. At this time the laws of chemical combination had been formulated and accepted—the laws, viz., which may be epitomised by saying that "chemical elements combine together only in the proportion of their equivalent weights, or in simple multiples of those proportions."

Dalton had propounded an explanation of these laws in his
" Atomic Theory," according to which chemical combination was due to the union of chemically indivisible particles,
the particle or atom of each element having its own particular fixed weight.

It was natural that other minds, impressed by Dalton's
theory, should seek for other such numerical relations in
the hope of fresh discoveries of Nature's laws.

In 1815 a paper appeared in Thomson's " Annals of
Philosophy," by Dr. Prout, in which he pointed out certain
apparent relations between the " atomic weights " of the
elements as then determined. The idea was at once taken
up by other chemists, and took shape in the following form,
known as *Prout's Hypothesis :* " The weight of the atom
of each element is a simple multiple of the weight of the
atom of hydrogen." The observed deviations were referred
to errors of experiment, just as the apparent deviations from
the laws of chemical combination were referred to experimental error.

It was the life-work of Stas to investigate both assumptions, and to show that while the laws of chemical combination are rigidly exact, the supposition of Prout is unsupported
by experimental evidence.

Prout's hypothesis owes its importance in the history of
science to the fact that it seemed to restore the old theory
of the unity of matter, which appeared to have received its
death-blow with the discovery of the chemical *elements*.
But if the atom of each element be exactly once, twice, or
thrice the weight of the atom of hydrogen, then it is
reasonable to suppose that the atoms of all elements contain
only one kind of matter, and that the hydrogen atoms are
the one class of ultimate particles of which all matter is
built up.

As the art of chemical analysis developed under the
hands of the great Swedish chemist Berzelius, it became
evident that Prout's hypothesis was not tenable in its
original form. It was revived, however, in a modified

shape, chiefly owing to the influence of Dumas. In the modified form, the hypothetical unit weight was that of the half-atom of hydrogen. Later on, Dumas was compelled to retreat yet further from the original position, and to take the quarter-atom of hydrogen as the greatest common divisor of the atomic weights. In this modified form the idea of Prout loses much of its interest, since the " quarter-atom" of hydrogen is itself an unknown thing. Nevertheless, the idea of the oneness of matter always exerts a certain fascination, and to some minds this unity of matter appears to be a logical necessity. Hence the tenacity with which the chemists have clung to the belief that apparent discrepancies were due to error of experiment, rather than to the inaccuracy of Prout's hypothesis.

Stas began his researches on atomic weights with a strong prepossession in favour of the hypothesis. He chose for his determinations such substances as could be prepared in a high state of chemical purity, and worked with large quantities of substance in order to eliminate the effect of errors in weighing. A large number of experiments, which occupied several years, furnished him with extremely accurate values for the relative weights of the atoms of silver, of the alkali metals, and of chlorine, bromine, iodine, and other elements. Moreover, the variety of methods employed served to eliminate possible systematic errors—errors, that is to say, not due to want of skill in the performance of an experiment, but due to the method itself. Each substance was prepared in several different ways and from different natural sources. Not the least remarkable tribute to Stas's skill is the remarkably close accordance between the values he obtained for the atomic weights by the various processes employed. The numbers obtained in his first series of researches were closely accordant among themselves, and wholly at variance with those demanded by Prout's hypothesis. Stas concludes his memoir thus: " Prout's hypothesis must be looked upon as a pure delusion ; the elements must be considered to be distinct entities, with no relation between their atomic weights."

The accuracy of Stas's work was admitted on all sides, but his conclusions were contested. The criticisms of the Genevese chemist, Marignac, are historically important, having led Stas to his second and more celebrated research.

Marignac contended that it was far from being proved that the constituent elements of many chemical compounds were present exactly in the proportion of their atomic weights. It was possible that many chemical compounds contained normally a very small excess of one or other of their constituents. This criticism strikes at the basis of the atomic theory, since that theory is founded on the assumption that the laws of chemical combination are mathematically exact. For half a century the scientific world had accepted the dictum that the laws of chemical combination were *lois mathématiques*, but the original experiments on which these laws were based were far from being models of accuracy.

This fact was admitted by Stas, who undertook the laborious task of a re-examination of those laws, with a view to settle by the most exact methods whether the laws were in fact of mathematical exactness, or, like so many physical "laws," only *lois limites*, or approximate relations.

In 1865, five years after the publication of his first series of researches, appeared the "Nouvelles Recherches sur les Lois des Proportions Chimiques." In this work Stas repeated the more important of his former determinations of atomic weights, with additional precautions. He also subjected to the most rigorous tests the laws of definite, constant, and equivalent proportions which had hitherto rested on the comparatively rough experiments of Dalton, Wollaston, and other workers of the early part of the present century. In this great work Stas confirmed, on the one hand, his previous conclusion that Prout's hypothesis was unsupported by experiment, but showed, on the other hand, that the laws of chemical combination, hitherto accepted on insufficient data, were, as far as experiment could prove, actual and veritable mathematical laws.

It is impossible to over-estimate the benefit conferred upon science by a man who has the courage to devote years of patient labour to the re-examination of supposed laws which have been accepted on the evidence of insufficient experimental data.

From the point of view of the working practical chemist, the most important aspect of Stas's researches is that relating to the preparation of chemical substances in a state of purity. Since Stas's time chemists have not been satisfied with the approximate purification of substances which in general sufficed the earlier experimenters. The approximate isolation or purification of substances is the first step in a chemical research; the complete purification is the most difficult and the most important part of exact research in the science. Stas's methods of purification have served as a model for all subsequent experimenters.

In order to give a general idea of the character of his work, we will describe a method he adopted for the purification of silver—a substance which is, as he says, the "pivot" of his determinations. Silver is a substance which, as Stas showed, can be obtained in a state of almost perfect purity. The way in which it resists oxidation, and the distinctive character and insolubility of certain of its salts, would lead one to suppose that its complete purification would be very readily effected. That this is not exactly the case will be evident from the following description of Stas's method. In order not to make the description unduly long, we omit the special methods of purifying the *reagents* used in the work. These reagents are water, nitric acid, hydrochloric acid, caustic potash, and milk sugar. Each of these had to be submitted to special processes of purification, lest their use should introduce foreign substances into the silver.

Coinage silver was taken, and dissolved in very dilute nitric acid. Any gold present is left undissolved. The solution of the nitrate is evaporated to dryness, and heated till no more nitrous fumes are evolved. The salt is then dissolved in a small quantity of water. On filtering, any platinum present is left behind. The filtrate is then diluted with about thirty times its volume of water, and an excess of hydrochloric acid added. All the silver is thus precipitated or thrown down in the form of the insoluble chloride of silver. Any copper and iron present remain in solution. The liquid is poured off, and the precipitate washed, first with dilute hydrochloric acid, and then with water, till the washing appears to be pure water containing no trace of copper or of hydrochloric acid. This washing of a large quantity of a precipitate is a very lengthy and tedious operation, requiring days or weeks, according to the quantity. The washing was effected in this case by shaking up the precipitate with water in a stoppered flask, allowing the precipitate to settle, and pouring off the liquid. All the operations with chloride of silver were carried out in a room lighted by artificial light, since daylight, as is well known, effects a chemical change in the composition of the substance. The chloride of silver, purified as above, is brought on to a cloth (previously washed with hydrochloric acid) and the water squeezed out. After drying, the chloride of silver is pounded finely in a mortar, and reduced to the metallic state by warming for forty-eight hours with a solution of caustic potash and milk sugar (both carefully purified). The finely divided metal is then fused, with special precautions to prevent access of impurities. By this process Stas hoped to obtain an ingot of perfectly pure silver, but found that, besides very slight traces of other substances, there remained an appreciable quantity, two parts in one hundred thousand, of silica.

Experience convinced Stas that no substance can be obtained absolutely pure, except by distillation. He therefore subjected the silver obtained as above to the process of distillation from one cavity to another in a hollowed block of quicklime, made from white marble. The cavity having been previously heated by the oxy-hydrogen flame, in order to drive off any volatile substances, such as soda, the silver was placed in the cavity and fused. No scum appeared on the surface, showing the absence of certain impurities, such as iron, which, under these circumstances, would form a slag. The heat from the oxy-hydrogen flame was then increased till the metal began to boil. The vapour had at first a strong yellow tinge, showing that sodium was still present.

This, however, soon disappeared, the vapour of the silver showing no colour, except a faint blue tinge. The absence of any green tint showed that the substance was free from copper. The metal having completely distilled into the second cavity, or receiver, in the lime block, it was found that absolutely no residue remained, the small quantity of silica, and any similar fixed substance of an acid character, having combined with the lime, and any oxidisable material having been burnt away by the flame of the oxy-hydrogen blow-pipe.

By the above process Stas believed that he had obtained silver absolutely pure, but Dumas subsequently showed that silver thus prepared absorbs, after distillation, but while still molten, a certain quantity of oxygen which does not combine chemically with the silver, but remains " occluded" in the metal. The elaborate precautions adopted by Stas were therefore not successful in obtaining even this well-known and characteristic substance in a state of perfect purity, though he subsequently determined the amount of oxygen present. But the practical chemist owes to Stas a proper appreciation of the difficulties attending the purification of substances, an appreciation of the necessity for taking every means to overcome these difficulties, and a knowledge of methods for the carrying out of this class of work; methods elaborated by Stas more than thirty years since, and which still form the basis of many of the recent researches on the determination of atomic weights.

CHAPTER V.

SECT. I.—MENDELÉEFF'S SYSTEM.

THE researches of Stas appeared to show that the connection between the atomic weights, which Prout thought he had discovered, was either unreal, or, at all events, not demonstrable.

Four years after the publication of Stas's second series of researches, the Russian chemist, Mendeléeff, made known his system of classifying elements on the basis of atomic weights—a system which has stood the test of experiment, and has pointed the way to many new paths of fruitful research.

Chemists had been for some time familiar with the fact that among the elements are certain "natural families," the members of which bear a general similarity to one another, and show a regular gradation of properties following the increase in atomic weight as we proceed from the lowest to the highest member. Thus chlorine, bromine, and iodine have certain properties in common, which mark them as members of a group or family. Among the members of this family the properties vary in a *continuous* manner with the atomic weight. Thus chlorine (at. wt. about 35½) is a gas, bromine (at. wt. about 80) is a liquid, and iodine (at. wt. about 126½) is a solid, and so in similar gradation with their other characters.

Lithium (7), sodium (23), and potassium (39), form

another natural family, the well-known alkali metals. Here, again, there is a continuous gradation of properties with rise of atomic weight, lithia, the oxide of lithium, being a weaker base than soda, and potash being the strongest base of the three oxides. The numbers attached to the names of the elements indicate how many times heavier their atom is than the atom of hydrogen.

The members of a natural family are termed *homologous* elements. The connection between the properties and the atomic weight among groups of homologous elements was, as we have said, known before the publication of Mendeléeff's first paper, which appeared in 1869. Mendeléeff, however, by comparing together the members of *different* groups or families, was led to the discovery of a new and peculiar relation between the weights of the atoms and their properties, a relation which is the basis of the present system of classification of the elements, known as the *Periodic System.*

When we write the elements* after hydrogen (1) in order of atomic weight—

Li. Be. B. C. N. O. F. Na. Mg. Al. Si. P. S. Cl. K. Ca.
7 9 11 12 14 16 19 23 24 27 28 31 32 $35\frac{1}{2}$ 39 40

and so on, there is found to be a regular gradation of properties with increase of atomic weight from lithium to fluorine, which are as opposite in their characters as any two elements with which we are acquainted. Increase of atomic weight from 7 to 19 has continuously diminished the electro-positive or metallic character possessed by lithium till we reach fluorine, a non-metal and the most strongly electro-negative element known.

After atomic weight 19 the gradation of properties does

* Approximate numbers are, for convenience, given for the atomic weights. The names of the elements symbolised above are : lithium (Li), beryllium (Be), boron (B), carbon (C), nitrogen (N), oxygen (O), fluorine (F), sodium (Na), magnesium (Mg), aluminium (Al), silicon (Si), phosphorus (P), sulphur (S), chlorine (Cl), potassium (K), and calcium (Ca). We cannot as yet render account of the positions of hydrogen, helium, and argon in this classification.

DMITRI IVANOWITSII MENDELÉEFF.

(Reproduced from a photograph by Warwick Brookes, of Manchester.)

not continue; on the contrary, there is a sudden "reversion to type" the next element, sodium (Na 23) being strongly metallic in character. It is, as we have already mentioned, a member of the same natural family as lithium, and in that family stands next to lithium in order of atomic weight.

As we proceed from sodium, in order of increasing atomic weight, we find once more a gradation from the most strongly marked metallic properties to the most decided non-metallic character. Magnesium (Mg 24) is a metal as sodium is, but its oxide is less strongly basic than soda. Alumina (the oxide of aluminium, Al 27) is weakly basic or weakly acidic, according to circumstances. Carbon, a non-metallic body, forms a weakly acidic oxide, and sulphur (S 32) is not metallic in its physical properties, and forms an oxide which is strongly acidic or acid-forming. The next element, chlorine (Cl 35½), is the first homologue of fluorine, and is a typical non-metallic element. The element next following, potassium (K 39), is, however, a metal, and forms a strongly basic oxide. It is the third member of the group of alkali metals; so that in passing from chlorine to potassium we have the second instance of "reversion to type."

After potassium, the metallic character again begins to decrease; the next element, calcium (Ca 40), forming an oxide, lime, which is a weaker base than potash. Thus, after potassium, as after sodium, the variation of properties goes on continuously with increase of atomic weight from one element to another for another *period*.

We will not follow these *periodic* variations further, partly because the relations become more intricate and more difficult to follow as we proceed to the higher atomic weights. We have gone far enough, however, to show the peculiar and novel character of the relation which Mendeléeff discovered. Whereas, the properties of the elements of any one family vary *continuously* with the atomic weight, the properties of all elements are a *periodic* function of the atomic weight; a certain increase of atomic weight being accompanied by a recurrence of certain properties possessed by an element lower in the scale.

This may be made clearer to the eye by writing the above list of sixteen elements in a somewhat different way. As there is a return to the metallic character in the seventh element from lithium, we will begin in our table a second line with sodium on the left, thus—

Li 7,   Be 9,   B 11,   C 12,   N 14,   O 16,   F 19,
Na 23,   Mg 24,   Al 27,   Si 28,   P 31,   S 32,   Cl 35½,
K 39,   Ca 40, etc.

We see that after a *period* including seven elements there begins a second period of seven.   The difference of atomic weights between the first and eighth element, between the second and ninth element, and so on, is in every case, as will be seen on examining the table, sixteen units or nearly so.   Two successive members of any family are separated by six intervening elements, and differ from one another by about sixteen units of atomic weight.   The addition of mass or weight to the chemical atom, from the mass 7 of the lithium atom to the mass 39 of the potassium atom, is not accompanied by a continual, unbroken increase of certain properties, but by a *periodic* variation of those properties.   The interval from lithium to potassium comprises two periods, each of which contains seven elements. The vertical rows contain the natural families (as lithium, sodium, potassium, and beryllium, magnesium, calcium). Previous to Mendeléeff's work relationships could only be clearly traced between members of the same family (or homologous elements), but the periodic system of classification enables us to trace the connection between the *heterologous* elements.   Among the following :—

Be 9,
Na 23,   Mg 24,   Al 27,
Ca 40,

sodium (Na), magnesium (Mg), and aliminium (Al), are termed heterologous elements, and beryllium (Be), magnesium (Mg), and calcium (Ca), homologous elements.

The elements sodium, aluminium, beryllium, calcium are termed the four *analogues* of magnesium. Mendeléeff showed that the properties of any element are completely determined by that of its four analogues. Thus, supposing the properties of the element magnesium were wholly unknown, they could be deduced from the properties of the analogues. Thus the atomic weight will be the mean of those of the four analogues. Now

$$\frac{9 + 40 + 23 + 27}{4} = 24\frac{3}{4}$$

which gives (approximately) the atomic weight of magnesium. Again, take the specific gravities. They are as follows :—

Sodium ...	...	...	specific gravity	0·97
Aluminium	...	...	,,	2·56
Beryllium	...	...	,,	2·10
Calcium...	...	...	,,	1·58

Mean of these values = 1·8.
Specific gravity of magnesium = 1·75.

It will be noticed that although the difference between the atomic weights of neighbouring elements in the horizontal rows (heterologous elements) is not absolutely constant, yet the variations are small, the interval rarely exceeding one or two units in the foregoing table, except in the case of the interval fluorine—sodium and chlorine—potassium.

At the time when Mendeléeff first drew up his table of the elements, it was found that in several cases the neighbouring heterologous elements did not fall into place—that is to say, did not come into the same vertical row with other members of the same natural family. Thus the element next to zinc (Zn 65), which belongs to the same family as magnesium and comes vertically below it, was arsenic

(As 75), which thus comes vertically below aluminium, although its properties are similar to those of phosphorus, not to those of aluminium.

Thus we had in the *second* and *fourth* horizontal rows—

	Na.	Mg.	Al.	Si.	P.	S.	Cl.
	23	24	·27	28	31	32	35½
and	Cu.	Zn.	As.	Se.	Br.		
	63	65	75	79	80		

Selenium (Se) and bromine (Br) have properties similar to those of sulphur and of chlorine respectively.   These facts led to an idea entirely novel in chemistry—that of *gaps* among the elements.   Hitherto the existence of an element with any particular atomic weight and particular properties had been regarded as an isolated and, so to speak, an accidental fact in Nature ; but Mendeléeff's generalisation introduces the idea of the necessity for the existence of elements with such and such atomic weights, and such and such properties.

It appeared extremely probable that there existed two elements intermediate in atomic weight between zinc and arsenic, between which there is an interval of ten units. Supposing two such elements to exist (called provisionally eka-aluminium and eka-silicon), Mendeléeff arranged the elements in the fourth horizontal row thus—

2nd Row—Na 23	Mg 24	Al 27	Si 28	P 31	S 32	Cl 35½
4th Row—Cu 63	Zn 65	Eka-Al	Eka-Si	As 75	Se 79	Br 80.

Reasoning from the assumption that the properties of an element are the mean of those of its four analogues, Mendeléeff drew up a table representing the properties of the hypothetical elements eka-aluminium and eka-silicon.

Two elements having the atomic weights required by the position of eka-aluminium and eka-silicon in the table have since been discovered, and named respectively gallium and germanium.   Their properties agree very closely with those predicted by Mendeléeff.

This power of prediction of hitherto unobserved elements

was an enormous stride in chemical science. The discovery of gallium holds in the history of chemistry a similar place to the discovery of Neptune in astronomy. The periodic system of classification enables us not merely to say with every confidence that such and such elements exist, though yet unobserved, but it puts us in a position to limit the number of possible, or at all events probable, elements. It enables us to predict with considerable accuracy the properties of chemical compounds before these compounds have been actually investigated, and it has in numberless ways proved of the greatest service to systematic chemistry. The philosophical interest of Mendeléeff's great generalisation is not inferior to that of the discovery of the laws of planetary motion. The Russian chemist, like Copernicus and Kepler, has shown the existence of law or order in one of the great departments of Nature's administration.

Newton showed that the laws of planetary motion discovered by Kepler were the necessary outcome of the property of universal gravitation. We yet await the discovery of a cause which will account for the connection between the weights or masses of the atoms and their properties. Such a discovery would be of surpassing interest.

Previous to Mendeléeff, all that was known of functions dependent on masses derived its origin from Galileo and Newton, and appeared to indicate that such functions always either increase or decrease with the increase of mass, as in the case of the attraction of celestial bodies. The numerical expression of the phenomena was always found to be proportional to the mass, and in no case was an increase in mass followed by a *recurrence* of properties such as is disclosed by the periodic law.

SECT. II.—ARGON.

(Written Feb., 1894.)

The Council of the Royal Society did well to hold an open night for hearing the important communication upon Argon, the newly discovered constituent of the atmosphere. The reading public is now acquainted with the principal properties of this peculiar gas, and with the course of patient investigation which led Lord Rayleigh and Professor Ramsay to its discovery. It is now time to inquire into the significance of the discovery, and the interest it possesses not only for scientific men, but for that larger body of people whose education in this country is still mainly literary, and not scientific.

The authors of the discovery have shown that the inactive residuum of the air which remains after removing oxygen and the smaller quantities of carbonic acid and aqueous vapour, does not consist wholly of the elementary gas nitrogen, but contains also in no small quantity another gas, which is also inert in its chemical behaviour, but which differs greatly from nitrogen in its specific gravity, being heavier in the proportion of 20 to 14. The volume of the new gas is more than $\frac{1}{100}$th part of the volume of the nitrogen; it appears to constitute, in fact, about one per cent. of the volume of the whole atmosphere.

The authors were led to this discovery by Lord Rayleigh's observation that nitrogen gas prepared from ammonium salts, nitrous oxide, or other chemical compounds was, bulk for bulk, half per cent. lighter than "atmospheric nitrogen," or that part of the atmosphere left after the removal of oxygen, carbonic acid and aqueous vapour. Which result, if either, was to be regarded as the true weight of the gaseous element nitrogen?

The lighter gas, prepared from chemical compounds, was found to be free from any admixture of hydrogen or other light gases which would lower its specific gravity. The electric discharge proved powerless to lower the specific gravity of the gas, as would be expected to happen if its lightness had been due to a partial splitting up of the molecules. The conclusion appeared to be that the greater weight of the gas derived from the air was the abnormal result, not the lower weight of the gas prepared from chemical compounds. In other words, the inactive residuum of the air contains something *heavier* than nitrogen gas.

It may well be asked how it could happen that the new gas has lain hidden from our ken till now, seeing that chemists have told us for more than a hundred years past that they had found air to be composed essentially of only *two* gases.

A ready answer, but not quite a sufficient one, is that argon is inert, and not apt to put itself in evidence by acts of chemical energy. In this respect its character is similar to that of nitrogen. "You can't see nitrogen," said a schoolboy; "it doesn't smell, and it won't explode. I call it a duffing gas." So is argon, only more so; but this is not the sole reason why it has been so long confounded with the nitrogen with which it is admixed.

More than a hundred years ago (1785), that careful worker, Henry Cavendish, expressly guarded himself against the assertion that the inactive part of the air (or, as he termed it, the phlogisticated part) was wholly composed of the gas which unites with the oxygen to form the acid of nitre. He says: "If there is any part of the phlogisticated air of our atmosphere which differs from the rest, and cannot be reduced to nitrous acid, we may safely conclude that it is not more than $\frac{1}{120}$th of the whole."

Unfortunately, the discovery of the nature of the chemical elements led to the *naming* of the phlogisticated part of the air. It was called nitrogen, and want of exact knowledge was hidden by precise nomenclature.

What Cavendish showed was that ninety-nine per cent. of the phlogisticated or inactive part of air consisted of the gas which combines with oxygen, under the influence of the electric spark, to form a substance which, in its turn, unites with potash to produce nitre. The remaining one per cent. he does not pretend to be certain about. One per cent. is sufficiently near for a first approximation; such results must be *provisionally* accepted if progress is

to be made, but as soon as experimental methods have
sufficiently improved the experiments should be repeated,
to see if the original results are the crude expression of a
loi mathématique or if they express a *loi de limite.*

Many of the results obtained by the earlier chemists
were subjected to revision (and for the most part were
confirmed) by the great Belgian chemist, Stas. Owing
to the difficulties of constructing the required apparatus.
Stas left unchecked some of the earlier experiments upon
nitrogen, the repetition of which might have led him to
the discovery of argon. Lord Rayleigh's well-known
work on the relative densities of the elementary gases
was intended, we believe, to check the conclusions at
which Stas arrived as to the relations between atomic
weights. This research has incidentally led him and
his coadjutor, Professor Ramsay, to revise a portion of
the work of the early chemists. It would be difficult
to mention any modern work in chemistry of greater
scientific value than the revision of first approximations
undertaken by Stas, and more recently by Raleigh and
Ramsay. In all such cases the critic is tempted to say
that the work of revision should have been undertaken
sooner, and that we ought not to have been so long
content with the first approximation. It has been well
said that it is in the investigation of residual phenomena
that important discoveries in the physical sciences are
nowadays most often made. There are newly discovered
"laws" in chemistry and chemical physics which have
only been shown to be true to a first approximation.
The discovery of argon may perhaps be an encourage-
ment to the more rigid investigation of the fundamental
experiments upon which these "laws" have been based.
But as travellers know, pioneering is more attractive
than surveying, and there is a deal of human nature
left in scientific man.

The authors have been successful in preparing fairly
large quantities of the new gas by repeatedly passing

air (freed from oxygen, moisture, and carbonic acid)
over strongly heated magnesium, until all the nitrogen is
taken up by the magnesium. The density of the pure
argon was determined by weighing it in a glass globe,
and the results obtained varied from 19·7 to 20. In
other words, if a given bulk of hydrogen weighs 1, the
same bulk of argon has a weight between 19.7 and 20,
the weight of the same bulk of nitrogen being almost
exactly 14.

What part does the element play in the economy of the
everyday world around us ? It is present in vast quantities,
one part in every hundred parts of the air we breathe,
gallons of it in every room. M. Jourdain talked prose all
his life without knowing it. We have breathed argon all
our lives and never knew it.

Do the molecules of argon remain for ever idle denizens
of the air, or do they, like the molecules of nitrogen, of
oxygen, and of carbonic acid, pass through a transmigration
of bodies, as constituents of minerals, plants, and animals?
It seems unlikely that the higher animals should have the
power of directly assimilating argon. Nitrogen we know
they cannot take up from the air.

In the mineral world one experiment has already been
tried with negative results. Professor Roberts Austin
calculates that 1000 cubic feet of argon are passed through
the molten metal in the charge of a Bessemer steel con-
verter. He has made the steel give up its combined or
occluded gas, but no argon was found.

The nitrogen of the air is with difficulty brought into
chemical combination. The electric spark, indeed, makes
it combine with oxygen slowly, and, as it were, reluctantly.
Argon, under these conditions, does not combine with
oxygen.

The chief agency by which nitrogen is brought into the
cycle of chemical combination and re-combination appears
to be the action of bacteria associated with the life processes of plants. And this is the age of bacteria. The

nimble microbe is a potent factor among the agencies
of science. Professor Ramsay appears to regard the new
element as a sort of chemical Topsey ; he " guesses nobody
can't do nothing with " argon. He has tried the violent
methods of heat and strong chemicals. Perhaps the
bacteriologist, with gentler methods, may yet show that
argon does not stand aloof from the ceaseless changes of
living forms around us, in which so active a part is played
by the other gases of the atmosphere.

SECT. III.—HELIUM.
(Written June, 1895.)

The search for new sources of argon has led to the
discovery of another element, which, like argon, is a gas.

The statement of Hillebrand, that "nitrogen" was given
off from the mineral uraninite when treated with dilute
sulphuric acid, called for re-examination in the light of the
recently discovered association of argon with atmospheric
nitrogen. It has now been found that some specimens of
uraninite do, in fact, give off nitrogen, but that the gas
usually obtained from uraninite and the allied uranium
minerals is neither nitrogen nor argon, but a new gas, to
which the name helium is given.

This gas, helium, has been investigated by Professor
Ramsay, who adopts the same methods as are used in the
case of argon to remove any possible trace of hydrogen,
oxygen, and other known gases. These processes (*e.g.*
passing electric sparks through the gas in presence of
potash) obviously would not separate it from argon, in fact
the method was adopted in the expectation that the gas
would prove to be argon itself. When, however, a tube
full of the gas was caused to glow by electricity, and the
light examined by the prism, the spectrum was not found
to be that of argon, but of some substance new to chemists.
The density of the gas was approximately determined, and
found to be little more than twice as great as that of
hydrogen, the lightest of known gases. The velocity of

sound in the gas was also with some difficulty ascertained ; with the important result that, as with argon, the "ratio of the two specific heats," which is connoted by the velocity of sound, corresponds to a *monatomic* constitution.

It appears therefore that the molecules, *i.e.* the smallest physical portions of the gas, are formed of only *one* chemical atom, hence the gas cannot be a chemical compound, but, like argon, is elementary. It may contain only one element, or it may be a mixture of two or more monatomic elementary gases. If it be a single element, the weight of the atom would be about four times the weight of the atom of hydrogen.

The spectrum of the new gas is brilliant and characteristic. One of the most remarkable lines is a strong yellow one which coincides, or (as Runge contends) *nearly* coincides, with a well-known line in the spectrum of the sun, which is marked D 3 on the map of the sun's spectrum.

Mr. Norman Lockyer long ago drew attention to this D 3 line in the sun's spectrum, which was not shown by any then known chemical substance. Believing that its presence was due to the existence of a new element, he termed it the "helium line." Helium was accepted as a provisional name for an element supposed, but not proved, to exist in the sun. It seems probable that Professor Ramsay's new gas contains this element, and the name helium has been given to it. Sometimes it is called "terrestrial" helium for the sake of avoiding assumptions. As usual, we have been in too great a hurry to name the new discovery. Mr. Norman Lockyer has been busy with the comparison of the spectral lines of the gases obtained from uranium minerals with hitherto unidentified lines in the sun and star spectra. He finds an apparent coincidence, not only in the case of the yellow line D 3, but in many other lines of the stellar spectra. He believes that we are "in the presence of a new order of gases of the highest importance to celestial chemistry." *

* Helium and argon have since been detected in meteoric iron.

G

On comparing the spectra of argon and of the new gas helium, it appears that although essentially different, yet two lines, at least, are common to both. This seems to indicate the presence in each gas of one or more elements common to both.

If this be so, there is a *third* element, and perhaps others, yet to be isolated from argon, and perhaps from helium.

As regards argon, this does not come altogether as a surprise. Probably the names argon and helium will eventually be retained for the two elements present in largest proportion in the new gases.

It is noteworthy that argon has only been found in the atmosphere, and helium only in minerals. In this connection a paper by Dr. Johnstone Stoney is of interest. Reasoning from the kinetic theory of gases, and regarding the molecules as projectiles subject to the influence of the earth's attraction, Dr. Stoney concludes that in gases much lighter than water-vapour (of which the density is nine times as great as that of hydrogen), the rapid motion of the molecules would ultimately carry them beyond the sphere of the earth's attraction. It should be premised that the average velocity of gas molecules is known, and that the velocity is greatest in the lighter gases. If a light gas such as helium, having a density only twice that of hydrogen, find its way at any time into the atmosphere, it will ultimately find its way *out* of the atmosphere. Those molecules which are not fixed terrestrially by chemical combination will find their way into space.

The moon, with its weak force of gravity, retains practically no atmosphere. The sun, on whose surface the force of gravitation is enormous, shows in its spectrum the hydrogen lines and a line which appears to characterize the light gas "terrestrial helium." In the stars, where the gravitating force is comparable to that on the sun, there appears to be some evidence of the presence of the new light gas. Hydrogen we know exists in their atmospheres. In our own atmosphere hydrogen only exists as water-vapour in which, heavily weighted by the attached oxygen atom, it is tied down to earth, something after the manner of a captive balloon. Free hydrogen would move off, like Jules Verne's projectile, to other spheres, not however from the earth to the moon, neither would it return, as Jules Verne's travellers did, to life terrestrial.

These considerations seem to afford a probable explanation of the singular circumstance that the existence of an element showing the D 3 line was first detected by the examination of a body ninety million miles away, although it is present in stones which have been known and handled for years.

In scientific research, as in everyday life, it is not always easiest to see what is closest to one's eyes. If the argon molecules had been active enough to escape from the atmosphere, and if those atoms had been collected and removed beyond the orbit of Mars, then the solid ball into which the gas might be frozen by the cold of space would be sufficient to form a very respectable asteroid, as asteroids go. It would not be one of the largest, but it would be a body by no means contemptible to the astronomers. It would be considerably bigger than the two moons of Mars rolled into one, and might have been discovered years ago by the telescope.

However, we have got hold of argon at last; also helium, and shortly there may be one or two more; a large proportional addition to the limited number of known chemical elements.

CHAPTER VI.

CHEMICAL SYMBOLISM AND ORGANIC CHEMISTRY.

SECT. I.—CHEMICAL SYMBOLISM.

THE symbols of Alchemy had a mystic meaning, and were connected with the symbols of Astrology. The Sun and the metal gold had the same symbol, so had iron and Mars, the planet Mercury, and the metal quicksilver. It was esteemed a merit of symbols that their significance was unknown and mysterious except to the initiated.

The re-introduction of symbols into chemistry along with the atomic theory was, on the other hand, intended to make the subject more readily comprehensible. The modern chemical symbols, and their combination in formulæ and equations, make the record of chemical changes more easy to follow for all who take the trouble to master the chemical alphabet. If the alphabet is unknown, the use of symbols proves a stumbling-block to the general reader when he takes up an ordinary chemical text-book. Every educated person should make himself acquainted with the chemical alphabet. On the other hand, it would be well if the use of symbols were confined to cases in which their employment is conducive to precision of thought, instead of perverting symbols into an inferior shorthand in order to save the trouble of writing the names of chemical substances

Dalton used the following symbols to denote the atoms of oxygen, hydrogen, nitrogen, and carbon—

O	⊙	⏀	⊕
Oxygen	Hydrogen	Nitrogen	Carbon.

Supposing the reacting particle (or molecule) of water to be composed of one atom of oxygen and one atom of hydrogen, he gave the formula to water—

$$\odot \quad O$$

In the light of our present knowledge that the molecule of water is composed of *two* atoms of hydrogen and one of oxygen, we should express the formula for water thus—

$$\odot \quad \odot \quad O$$

in Dalton's symbols.

At the time these symbols were introduced, chemists had no information as to the *arrangement* of atoms in the molecules. As the above formula is only intended to express composition, not arrangement, it may be figured indifferently either as above or thus :—

$$\odot \quad O \quad \odot$$

or thus—

$$\odot \quad \quad \odot$$
$$O$$

or in any other manner in which the proximity of the symbols indicates that they are to be read together as constituting the formula of a complex molecule, or of the body which is composed of these molecules.

Hydrogen gas is composed of two-atom molecules. We should therefore express the formula for this substance thus—

$$\odot \quad \odot$$

in Dalton's symbols. Dalton was not aware that the molecule or ultimate particle of hydrogen consisted of more than one chemical atom. He would therefore have expressed the formula for the substance hydrogen by the use of the symbol for the atom.

"It was this happy idea of representing the atoms and the constitution of bodies by symbols," says Thomson, the historian of chemistry, "that gave Mr. Dalton's opinions so much clearness."

Berzelius introduced the present system of symbols for the atoms, namely, initial capital letters, followed, where necessary, by a succeeding letter of the name of the element, written small. The following are examples :—

C   Cl   Cs   Ce

Carbon   Chlorine   Cæsium   Cerium.

The name for which the letter stands must be the one by which the element in question can be recognised in any language, as the system of symbols is international. Many of the elements have different names in different languages, names which do not always commence with the same letter. The Latin or Greek name provides the initial letter for these, as the following examples show :—

Cu,   Cuprum,   Copper,   Cuivre,   Kupfer.
N,   Nitrogen,   Nitrogen,   Azote,   Stickstoff.

The initial letters are placed close together as formulæ for the molecules of substances. When there is more than one atom of an element in the molecule, a numeral is placed below and on the right, thus—

H_2 H_2O $H_2S\ O_4$

Hydrogen gas   Water   Sulphuric acid.

The use of the symbol H to denote hydrogen gas is a slovenly and pernicious practice. H properly represents the *atom* of hydrogen. H_2 represents the molecule of hydrogen, and its use may be extended so as to denote hydrogen gas, which is composed of particles or molecules, each of which has the composition H_2.

When the molecular weight of a substance is unknown, the *empirical* formula, expressing in the simplest manner the proportions between the number of atoms present, is sometimes employed. Thus starch is usually represented by the empirical formula $C_{12}\ H_{22}\ O_{11}$. As we only know that the molecular formula is a whole multiple of this, we may write $(C_{12} H_{22} O_{11})_n$, where n is an integer, as a molecular formula for starch.

The above formulæ merely express the percentage composition, and, when known, the molecular weights.

The next extension of chemical symbolism was the representation of the modes of formation or decomposition of substances in their formulæ. Thus the percentage composition of caustic soda is represented by the empiric formula Na OH. Caustic soda can be produced by the action of the metal sodium upon water (H_2O), when hydrogen is evolved, the atom of sodium replacing the atom of hydrogen in the water-molecule. When caustic soda is heated with sodium, the remainder of the hydrogen is replaced, oxide of sodium (empiric formula Na_2O) being formed. These relations are to some extent indicated by the formulæ

$$O \left\{ \begin{array}{l} H \\ H \end{array} \right. \qquad O \left\{ \begin{array}{l} Na \\ H \end{array} \right. \qquad O \left\{ \begin{array}{l} Na \\ Na \end{array} \right.$$

$$\text{Water} \qquad\qquad \text{Caustic soda} \qquad \text{Sodium oxide.}$$

These were called *typical* formulæ. They are a transition between the *composition* formulæ (empiric and molecular) and *constitutional* formulæ which were the next development of chemical symbolism.

The atoms of the different elements are not mutually *equivalent* in chemical change. An atom of oxygen seizes upon *two* atoms of hydrogen when water is formed, and is combined with two atoms of chlorine in chlorine mon-oxide. Chlorine and hydrogen combine atom for atom. The nitrogen atom is combined with three atoms of hydrogen in ammonia, and the carbon atom is combined with four atoms of hydrogen in marsh gas. Sodium replaces hydrogen atom for atom when sodium acts upon water. One atom of zinc replaces two atoms of hydrogen when zinc dissolves in dilute sulphuric acid. The chlorine and hydrogen atoms are equivalent to one another. This is expressed by saying that the chlorine atom is *monovalent*, and chlorine is called a monovalent element. The atoms of oxygen and zinc are equivalent to two atoms of hydrogen (and of chlorine), and oxygen and zinc are termed *divalent*. Nitrogen is trivalent, carbon tetravalent.

In the chemical transformations of substances it is observed that the polyvalent elements appear to bind together the parts of the molecule. Sometimes a polyvalent atom seems to have the power of keeping together a group of atoms which pass unchanged from one molecule to another in chemical reactions. Such groups are called *radicles*. Lines of attachment drawn so as to join the atomic symbols in the formula express these relationships in the graphic, constitutional, formula. Thus the constitutional formulæ for water, caustic soda, and sodium oxide are—

H-O-H	Na-O-H	Na-O-Na
Water	Caustic soda	Sodium oxide.

In these substances oxygen appears to be the binding atom, and this is symbolised by the lines of attachment. Without asserting that there is no direct attachment, or attraction, between the atoms of sodium and hydrogen in caustic soda, it is evident that the union between oxygen and sodium and oxygen and hydrogen, is much firmer and more intimate. The formula expresses this salient fact.

When hydrochloric acid (HCl) acts upon caustic soda (NaOH) we get sodium chloride (NaCl) and water (H_2O). This may be expressed by a *chemical equation*, thus—

$$NaOH + HCl = NaCl + H_2O$$

or, using constitutional formulæ,

$$Na - O - H + H - Cl = Na - Cl + H - O - H$$

Chemical equations epitomise our knowledge of *one-half* of chemical science; they express one of the two aspects from which substance-transformation may be considered. The left-hand side of the equation represents the composition of the substances present before they react. The right-hand side represents the composition of the sub-stances formed during the reaction. The equation records the substance-transformation which takes place during a chemical reaction. This is one-half of chemical science. The other half is the study of the conditions which bring about a reaction, and of the phemomena which accompany substance-transformation. The symbol + on the left-hand side of the equation stands for the conditions of reaction, the symbol = stands for the accompanying phenomena. There has been no develop-ment of this part of chemical symbolism corresponding to the development of molecular formulæ. As we are only dealing with the developments which have taken place in chemical symbolism, we shall have nothing further to say in this essay about the bald and barren symbols + and =.

To return to the constitutional formulæ. The action of sodium upon water, expelling half the hydrogen and forming an hydroxide, is one of a number of similar reactions. The hydroxides, containing an atom of a metal together with oxygen and hydrogen, are a numerous class of substances. The empiric formulæ for hydroxides containing the mono-valent, divalent, and trivalent metals, sodium, calcium, and aluminium, are—

$$NaOH, \quad CaO_2H_2, \quad AlO_3H_3$$

The corresponding constitutional formulæ are—

$$Na - O - H, \quad Ca \begin{matrix} \diagup O - H \\ \diagdown O - H \end{matrix}, \quad Al \begin{matrix} \diagup O - H \\ - O - H \\ \diagdown O - H \end{matrix}$$

Caustic soda   Slaked lime   Aluminium hydroxide.

The group of atoms or radicle (OH) which appears in hydroxides, and in a large number of other substances, acts in some ways like a monovalent atom. It is not really a chemical atom, as it *can* be chemically divided. In a large number of chemical reactions, however, it *is not* divided, and it is convenient and useful to bracket the two atoms together thus (OH), and, since the group is monovalent, to write them with one external line of attachment, thus—

$$- (OH)$$

We may then write the constitutional formulæ of the above hydroxides thus—

$$Na—(OH) \qquad Ca=(OH)_2 \qquad Al \equiv (OH)_3$$

or, *recollecting that* (OH) *is monovalent,* we may write the formulæ still more shortly thus—

$$Na(OH) \qquad Ca(OH)_2 \qquad Al(OH)_3$$

The condensing of cumbrous expressions is entirely analogous to the practice in algebra.

As the study of the atomic constitution of substances was elaborated, it was found that graphic formulæ written on the page of a book without regard to the circumstance that the atoms are not necessarily all lying in one plane, could not always be manipulated to represent beyond a certain point the chemical constitution and behaviour of substances. In such cases *glyptic* formulæ are employed, which in the form of models or of perspective drawings attempt to symbolise the *arrangement of atoms* in space of three dimensions. This, the latest development of chemical symbolism, arose from the detailed study of organic chemistry, and is referred to in the next essay.

The atoms are probably not relatively at rest within the molecule. Symbols representing fixed positions must therefore necessarily be imperfect, but a symbol which was other than imperfect would cease to be merely the symbol, it would be the thing itself.

Sect. II.—The Development of Organic Chemistry.

Organic chemistry, the study of the hydro-carbons and their derivatives, is a science of the present century. Some of the technical processes connected with the preparation of organic substances are, no doubt, of very ancient origin—

brewing, for instance, and the art of soap-making, but it is the scientific aspect of the subject with which we are here concerned.

Lavoisier showed that the decomposition of sugar by fermentation proceeds according to the conditions of the law of conservation of mass, the carbonic acid and alcohol produced being equal in weight to the sugar from which they are formed. Early in the present century the Swedish chemist, Berzelius, showed that the composition of organic substances conforms to the laws of constant proportions and of multiple proportions, which Dalton and others had shown to be characteristic of mineral compounds.

The way was now paved for the recognition of the study of organic materials as a part of the domain of scientific chemistry, conforming to the same laws as those which govern the chemical properties of mineral substances. Research in the organic branch of chemistry was immensely facilitated by Liebig's work in perfecting the principal process of organic analysis, the well-known "combustion" which is still the *pons asinorum* of the student's laboratory course. From Liebig's time the progress of organic chemistry has been marvellously rapid. The compounds of carbon are, for the most part, so "reactive" that the labour of the investigator is quickly rewarded by the production of some novel substance—often useful or curious—the discovery of which leads in its turn to the production of other bodies related to it.

The binding element in the majority of these substances is carbon. The carbon atoms seem to have an almost unlimited capacity for catching hold of and hanging on to one another, and at the same time they retain their hold upon one or more atoms of other elements with which they have been associated. Thus, in the laboratory of the plant or animal body, and in the laboratory of the chemist, are built up compounds of almost infinite complexity, though containing for the most part but few of the chemical elements. Carbon is present in all, hydrogen in almost all, and oxygen in the majority of cases. Nitrogen occurs frequently, and the other elements in smaller quantity and comparatively seldom. The known hydro-carbons—*i.e.* compounds containing only hydrogen and carbon—number four hundred, whilst few of the elements except carbon combine in more than two or three proportions with hydrogen.

The total number of carbon compounds is said greatly to exceed that of all the other known chemical substances. Among the organic bodies which have been produced in the laboratory are useful drugs and welcome anæsthetics; dyes, of which many are brilliant and some are beautiful; and powerful explosives, the discovery of which has benefited their manufacturers.

The facility with which chemists can transmute one carbon compound into another has led to great developments in our knowledge of the mechanism of chemical reactions, and of the chemical structure of molecules.

Chemical formulæ, from expressing merely the quantitative composition of compounds, were soon used to express the methods of formation and decomposition of organic substances. As knowledge advanced, it was seen that the formulæ could be made to indicate the way in which the atoms are united to one another. It appeared from the study of organic chemistry that the attraction or union of an atom is not so much with all the rest of the molecule as with some neighbouring atom with which it is closely united or related. The *graphic formulæ*, with which modern chemical books are filled, express symbolically the order or arrangement in which the atoms of the compound molecule are bound or linked together. To such perfection has the symbolical expression of the constitution of organic substances been brought, that the manipulation of these symbols often furnishes a valuable guide in the prosecution of new researches.

No mode of expressing graphically *on paper* the composition of a molecule can, however, be expected to be quite satisfactory if it fails to take account of the fact that the atoms of a molecule are not all distributed, and do not all move, in one plane. The ordinary graphic formula of the text-book has the same fault as a Chinese battle picture —it takes no account of perspective.

The more recent use of *glyptic* symbols (which look like outlined figures of crystal form), or of actual models, is an important extension of this domain of scientific symbolism.

We must explain shortly how these developments have come about. The study of carbon compounds led to the discovery of *isomeric* bodies which differ in their properties, although their analytical composition is identical. These differences must, it seems, be explained on the supposition of a different grouping of the chemical atoms, and the phenomena of isomerism are to be classed along with those of allotropy, which are exhibited by several elementary substances, notably by carbon itself. In the case of carbon compounds, it was found possible in many cases to express these differences by the graphic formulæ referred to above. Thus, there are two substances of very different properties, the percentage composition and molecular weight of which is expressed by the formula $C_2H_4Cl_2$—the symbols C, H, and Cl standing for the atoms of carbon, hydrogen, and chlorine. It was found that in one of the two substances whose composition is expressed as above, both chlorine atoms were bound up to the same carbon atom, whereas in the other the reactions showed that each carbon atom was in intimate connection with only one atom of chlorine. These facts are symbolised as follows :—

$$
\begin{array}{ccc}
& \text{H} & \text{H} \\
& | & | \\
\text{H}-\text{C}-\text{H} & & \text{H}-\text{C}-\text{Cl} \\
& | & | \\
\text{Cl}-\text{C}-\text{Cl} \quad \text{and} \quad & \text{H}-\text{C}-\text{Cl} \\
& | & | \\
& \text{H} & \text{H}
\end{array}
$$

But these graphic symbols are insufficient to explain some cases of isomerism. For such cases it is useful, instead of using the symbol —$\overset{|}{\underset{|}{C}}$—, to represent the carbon atom by a tetrahedron.

This symbol expresses the essential fact that the carbon atom has a fourfold power of union with other atoms, and we symbolically express the union by attachment to the *corners* of the figure. It is generally sufficient for the purpose in view if one or two carbon atoms are thus fully represented, the ordinary symbol sufficing to show the functions of the other carbon atoms.

The two substances, fumaric acid and maleic acid, have the molecular composition $C_4H_4O_4$, and the study of their reactions shows that each has two groups (CO_2H), in which the hydrogen is bound to carbon only through the connecting links of the oxygen atoms, whereas the remaining two hydrogen atoms are united directly to carbon. Two graphic representations may be made, which in a condensed form appear thus—

$$CO_2H \, . \, CH \, . \, CH \, . \, CO_2H$$

and $\quad CH_2 \quad . \quad C \quad < {\displaystyle {CO_2H \atop CO_2H}}$

The latter symbol would indicate that *two* (CO_2H) groups are united to one carbon atom. The reactions of the substance so represented are, however, altogether against this supposition, and the graphic representation therefore fails.

If, however, we represent the functions of two of the carbon atoms more fully by symbolising them as tetrahedra, we can represent very well the observed differences in the two acids, by taking account of the arrangement in the solid instead of regarding the apparent arrangement on the flat. In the symbols given below, the dotted lines of the tetrahedra indicate those edges which would be invisible if we were dealing with a wooden model, or with an opaque crystal of the same form. A class of bodies which is difficult to represent by any symbol is that known as *tautomeric*, in which one or more atoms appear to be in a state of alternating allegiance towards their more powerful neighbours. The hydrogen atom,

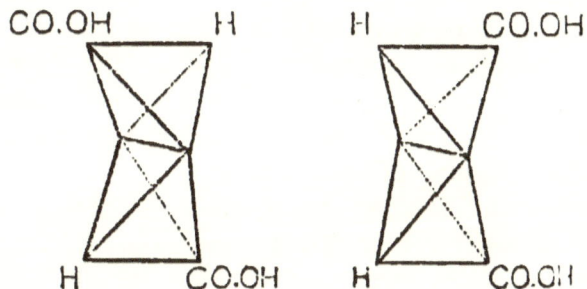

which is the lightest, and probably the fastest mover in the dance of atomic vibration, seems, in some cases, to have certain peculiar privileges of motion within the molecule, so that at successive moments it may be joined up to different atoms.

In the work of synthesising or building up compounds which occur in Nature, the organic chemists soon outstripped their " inorganic " brethren. Lately, however, the French representatives of inorganic chemistry have again recovered the lead, and the diamond has been made, while starch and the albuminoids are still beyond the creative powers of the chemist.

The early feats of organic synthesis were hailed as a triumph over the old belief in a mysterious *vital force*. All substances produced in the life-processes of animals and plants had been supposed to owe their existence to the mysterious agencies of life, and to be inimitable by the chemic art. The synthesis of a host of such substances in the laboratory has shown that the mysterious powers of vitality were prematurely invoked, and that failure was pre-supposed when no true trial had been made. The case is somewhat analogous to the mistaken appeal to almost infinite periods of time as the condition of formation of the native crystals of ruby and diamond.   In spite of all their past achievements, scientific men are ready enough, like other mortals, to cry out that their go-cart cannot get any further without the aid of some Herculean agency beyond their reach.

As a matter of fact, however, the achievements of organic synthesis have only pushed a *vital-force* theory a small distance further back, for none of the reproduced alkaloids, sugars, dyes, etc., are *organised* bodies, or show any sign or symptom of the germ of living power.

Recent researches upon the molecular weights of organic substances (chiefly by Raoult's method, which is based upon the lowering of the freezing point of solvents), appear to show that the simplest among the substances which are intimately associated with vital processes are of vastly higher molecular weight, and presumably vastly more complicated than any of the substances that have yet been synthesised.   The great differences in the size of molecules is perhaps indicated by the phenomena of dialysis, so much used in physiological work for the separation of substances. Crystalloids, sugar for instance, will, in solution, pass through the pores of an animal membrane, such as parchment, whereas *colloid* substances will not.   It seems likely, that in such substances, perhaps through the action of atoms such as those of carbon which have a power of multiplex combination, molecules or groups of atoms may "combine to net-work or sponge-like masses. . . . We may perhaps further suppose that through the constant change of position of polivalent atoms, these mass-molecules will show a constant change in the connected individuals, so that the whole . . . is in a sort of living state" (Schorlemmer's "Rise and Development of Organic Chemistry," p. 261, of Prof. Smithells' new and enlarged edition).   The idea thus brought forward, may, perhaps, be expressed by saying that if ever chemists should succeed in obtaining albuminous bodies artificially, it will be in the state of protoplasm. Now that an independent cell-life in the organism has been recognised, the organic chemist may be expected to make a serious attempt to ascertain if living matter can be produced otherwise than by the agency of living matter itself. Not every scientific man would be able to approach this world-old question without a preconceived opinion as to the ultimate answer.

Whatever be the answer which Nature has in store for us, it will be a duty to science to work at the problem until it is either solved in the affirmative, or, like the transmutation of metals, found by experience to be beyond our power.

Hitherto, chemistry has not been in a position to attack the problem. The synthesis of organic compounds must be carried still further before science will have a bridge long enough to span the wide and formidable gap which divides our knowledge of the inanimate from that of the living world.

CHAPTER VII.

CHEMICAL ACTIVITY.

SECT. I.—THE MEASUREMENT OF CHEMICAL ACTIVITY.

SOME substances are more potent than others in producing those substance-transformations which we call chemical changes. We say they are " chemically active." We know little of the causes of chemical activity, but we know how to compare quantitatively the activities of many chemical substances. It is possible, for instance, to draw up a table of numbers expressing the relative activities of acids, just as we can draw up a table of the relative weights of acids which are of equal combining value.

Tables of combining, or equivalent, proportions were drawn up before the atomic theory had explained the necessity for fixed and constant proportions of combination. The determination of the relative weights of equivalent quantities furnished an important part of the foundation on which was raised Dalton's theory of chemical atoms possessed of different specific weights. Similarly, the quantitative comparison of chemical activities is a part of the foundation on which must be based an explanation of the capability for combination which the atoms possess.

In order to compare the *chemical* activities of substances it is necessary to eliminate all influences not strictly chemical which affect substance-transformation. For example, the reaction which takes place on adding a solution of hydrochloric acid to a solution of silver nitrate is not suitable for comparing the chemical activities of nitric and hydrochloric acids, because the silver chloride, being insoluble in water, is removed from the solution as quickly as it is formed, and the nitric acid has scarcely any opportunity of getting at the substance. The two acids are not competing for the silver upon equal terms.

In order to compare the activities of the two acids we ought to select such a reaction as that of a solution of hydrochloric acid upon a solution of sodium nitrate. In this case as all the possible products are soluble in water, the nitric acid set free from the sodium nitrate has a fair opportunity of decomposing the sodium chloride. These considerations indicate how it is possible to arrange matters so that the hydrochloric and nitric acids shall compete upon equal terms for the base soda.

Let us mix in dilute aqueous solution the base soda and the two acids hydrochloric and nitric. This is more satisfactory than commencing with one salt already formed, since both acids have to start from the beginning. We must, however, have regard to the relative quantities employed. Perhaps an undue preponderance in the quantity of one acid will give it an advantage, not due to its superior activity, in the competition for the base. We have, therefore, to decide what weights of the acids must be taken in order that the competition may be strictly fair. If we were dealing with a merely mechanical competition we should choose *equal* weights; but this is a strictly chemical competition in which we have to compare the amounts of substance-transformation. We know that the quantities of substances which are of equal combining value are by no means physically equal in respect of their mass or weight. We must start therefore with *equivalent*, not with equal, weights of the two acids in order to secure a level chemical contest. As regards the quantity of the soda, it would not seem to affect the fairness of the competition whether the quantity taken be large or small; but it is convenient to the investigator so to choose the quantity of the base that the reaction shall be as simple as possible. Obviously it will greatly simplify the investigation if the quantity of the base be such as is just sufficient to saturate the acids, so that at the end of the reaction there will only be two substances in solution, sodium nitrate and sodium chloride.

To compare the activities of the two acids, it now only remains to ascertain in what proportion the base divides itself between them. If each acid obtains one half of the base the activities of the two acids are equal. If one acid obtains twice as much of the base as the other, then the activities of the acids are as 2 : 1.

The methods of determining the distribution are not the ordinary methods of chemical analysis, for in this case we are debarred from splitting up or analysing anything; we have to examine the system as it stands without introducing any disturbance of the balance. Physical methods are therefore employed. Knowing the physical constants of the substances, *e.g.* their specific gravity, refractive power, etc., these quantities are determined for the mixture and the proportion of the ingredients is calculated accordingly.

The result, in the above example, shows that hydrochloric and nitric acids are of practically equal activity, and the same result is obtained if we set them to compete for other bases, as potash or magnesia.

If we set nitric acid to compete with sulphuric acid, we find that the nitric acid is more active than the sulphuric in the ratio of about 2 : 1. Similarly, hydrochloric acid is found to be more active than sulphuric in the *same* ratio. We must take care, in arranging these experiments, that we make no mistake about the weight of sulphuric acid which may fairly be put to compete with one " equivalent " of hydrochloric or nitric acid. The ordinary chemical formula for hydrochloric acid is HCl, weight $36\frac{1}{2}$, and for sulphuric acid is H_2SO_4, weight 98. But an atom of sulphur is equivalent to *two* atoms of hydrogen, and the formula weight of sulphuric acid is *twice* its equivalent weight. For every $36\frac{1}{2}$ parts by weight of hydrochloric acid we must therefore use 49 of sulphuric acid, not 98, in order to ensure a level chemical competition.

The activities of acetic acid, of phosphoric acid, and of many other acids, are much less than the activities of the three common acids, hydrochloric, nitric, and sulphuric.

Another way of comparing the activities of two substances is to set them successively to perform the same chemical operation, and to observe how much more *quickly* one acts than the other. For example, a solution of cane sugar in water is broken up by the influence of acids into the two sugars called dextrose and lœvulose. The amount of transformation can be observed at any moment by noting the extent to which a beam of polarised light is twisted in passing through the solution. From the amounts of transformation effected in equal times we can deduce the relative activities of the different acids.

We may also employ the method for. comparing the rate of transformation effected by the same acid at different concentrations. It is found that *if the quantity of acid in a given volume of the liquid be increased, the rate of substance-transformation effected by the acid is increased in the same proportion.* This result, which is confirmed by other reactions, is important both to a theory of chemical activity and to practical chemistry. It is called, inappropriately, the Law of Mass, or the Law of Action of Mass. The name would appear to imply that the increased chemical effect of increased quantity is brought about by a stronger mechanical pull due to the greater attraction of a larger quantity of matter. This is not what the law implies; and we should not be justified in making such an assumption. If we double the concentration of the active substance, we double not only the mass present in a given volume but also the *number* of active particles, which from our present point of view is a much more important consideration. That mere *mass* has comparatively little to do with the progress of a reaction is evident from the cases in which a solid substance is formed during the reaction of liquids ; the condensed product taking scarcely any further part in the reaction. A more important condition than mass is the freedom of the particles to move about so that they have *frequent opportunities* of effecting chemical transformation.

Let us here recur to the reaction between a solution of hydrochloric acid and a solution of sodium nitrate.

The hydrochloric acid removes some of the soda from its combination, forming some sodium chloride and an equivalent quantity of free nitric acid. After a time no further change can be detected, and we have then a mixture of constant composition containing nitric acid, sodium nitrate, sodium chloride, and hydrochloric acid. The proportions in which these substances are present, depend first upon the relative activities of the nitric and hydrochloric acids, and secondly upon the relative number of equivalents of hydrochloric acid and of sodium nitrate which were taken for the experiment. If it were not for the nitric acid set free, the hydrochloric acid might complete the conversion of the nitrate into chloride. Free nitric acid, however, can form sodium nitrate by acting upon sodium chloride. We must suppose that when the two changes take place at an equal rate, the condition is reached, at which no further chemical change is *observed*. We have no reason to suppose, however, that either action *ceases*. Both probably continue, but at equal rates, and, as we cannot observe the individual actions between the ultimate particles but only the total effect, we observe *nothing*, because the effects are equal and opposite.

The reactions in which substances are removed, at the moment of their formation, from the sphere of chemical action, are the most suitable for the purpose of chemical analysis, and are as far as possible employed in the preparation of chemical substances, both in the laboratory and upon the manufacturing scale. Thus barium sulphate may be prepared from a solution of barium chloride, by adding *the equivalent quantity* of sulphuric acid, *the whole* of the barium being thrown out of its combination with hydrochloric acid in the form of insoluble barium sulphate.

Equally effective is the formation and removal of a gaseous product when liquids, or liquids and solids, interact. Thus calcium carbonate is completely converted to calcium chloride, by the action of the equivalent quantity of hydrochloric acid in solution.

In other cases where the products of the first action have the opportunity of reacting, it is necessary to employ more than the equivalent quantity of the agent whose action is to prevail.

An *excess* must also be used where the physical state of the substances is such as to hinder the reaction from progressing in the direction which the experimenter desires. Thus sulphuric acid is less volatile than nitric acid, consequently if a sulphate be heated with nitric acid it requires *a very large excess* of the nitric acid to expel all the sulphuric acid; although, as we have seen, nitric acid easily beats sulphuric acid when they compete for a base *upon equal terms.*

Thus the observed potency of chemical agents depends not only upon their specific chemical activity, but also upon the physical states of substances, a circumstance which much increases the difficulty of a scientific investigation of chemical activity.

SECT. II.—THE CHEMICAL ATTRACTION OF ATOMS.

The ultimate source of all chemical activity may be supposed to be a force by which the chemical combination of atoms is brought about. The chemical attraction of atoms may be called Chemical Force, a term intended to indicate its origin, not to express any special character of the force itself.

The capacity which atoms possess of joining up to each other in chemical union is one of the greatest of the world's sources of available energy. The general tendency of chemical action is towards those substance-transformations which are accompanied by evolution of heat. The heat evolved in chemical actions is the principal source of mechanical energy employed by man; wind and water being the chief exceptions. The power of man and other animals to perform mechanical work is also due to the slow combustion of food.

When $4\frac{1}{2}$ grams of water are formed by the interaction of molecules of hydrogen and oxygen gas, the heat given out is 17,100 *calories*, of which the mechanical equivalent is 724,185,000 gram centimetres, the amount of work required to raise 724,185,000 grams through 1 cm. against the force of gravitation acting between this mass of matter and the earth. The amount of work done is large in proportion to the amount of substance transformed.

In order to estimate the intensity of the force concerned in the substance-transformation, we require to know further what is the distance through which the force acts. The behaviour of gases shows that two molecules at a distance of ·00001 cm. (10^{-5} cm.) are, practically, outside each other's sphere of influence. In the course of their movements the molecules of gases sometimes come into close proximity, undergoing what is termed an "encounter." It appears to be during the encounter of molecules that an exchange of atoms can take place. The small distance between molecules during an encounter, must be supposed to be of the same order of magnitude as the diameter of a molecule, since the atoms during an encounter can pass over from one molecule to another.

Various considerations enable us to calculate the order of magnitude of the molecules. Thus, considering the average length of the free path of the molecule of a gas, and assuming that when the gas is reduced to a liquid the molecules nearly fill the whole space, we get the value ·00000001 cm. (10^{-8} cm.) for the diameter of a molecule. Other methods give concordant results, the values ranging from 10^{-7} cm. to 10^{-8} cm.

If we take 10^{-5} cm. as the maximum distance at which the chemical attraction between atoms in different molecules comes into play (*which is an outside estimate*), we see that the chemical attraction acts through a distance of *at most* 10^{-5}—10^{-8} cm. or say roughly 10^{-5} cm. Applying this result to the formation of water in which $4\frac{1}{2}$ grams were produced with a development of 724,185,000 *gram centimetres* of work, I calculate the average force to be about 10^{15} dynes. Taking Professor Boys' value $6\frac{1}{2} \times 10^{-8}$ dynes as the force of gravitation between two 1-gram masses 1 cm. apart, I find that the *gravitation* of $\frac{1}{2}$ gram of hydrogen and 4 grams oxygen at the distance required in the above example would be about 10^6 dynes. Gravitation, therefore, would only account for $\frac{10^6}{10^{15}} = 10^{-9}$, or one-thousand millionth part of the force brought into play in the above chemical reaction. We conclude, therefore, that in the chemical union of atoms a force other than gravitation comes into play.*

It is very difficult to unravel the phenomena of chemical attraction, on account of the fact that scarcely any of the substance-transformations of which we are able to investigate the energy-change, are simply reactions between individual atoms. In almost every case the atoms are joined up in molecules before the reaction, and in other molecules after the reaction. The attractive forces which come into play are, therefore, only the resultants of opposing forces. The development of energy which takes place

---

* Lord Kelvin supposes that at very close quarters the attraction of gravitation may be much greater than that calculated from the known law of inverse squares, which has only been proved to hold for distances from, say, 1 cm. upwards. He considers that cohesion may in this way be accounted for. Cohesion in solids, as far as I have been able to calculate its value, is of the order of magnitude of the forces which come into play in chemical action. If, therefore, gravitation force can really account for cohesion, it might be sufficiently *strong* to account for the chemical action of atoms, but chemical attraction appears to be a *directed* force, and to be capable of being more or less "satisfied."

when hydrochloric acid gas is produced by the reaction of hydrogen and chlorine, does not enable us to calculate the absolute value of the attraction of two atoms of hydrogen upon two atoms of chlorine, but only the excess of this attraction over the sum of the attractions between two atoms of hydrogen and between two atoms of chlorine.

Substance-transformation, in its ultimate character, is an exchange of constituent atoms between molecules, but we cannot *observe* substance-transformation in its ultimate character, a chemical phenomenon as perceived by us being merely the sum total of the effects produced by the interaction of vast numbers of molecules. Clerk Maxwell calculates, from the dimensions of molecules as approximately determined, that the smallest spec visible with a powerful microscope, contains something like two million molecules.

A knowledge of the attractive forces between the different molecules would not of itself be sufficient to explain the phenomena of substance-transformation as observed in the quantities of matter with which we can directly deal. In these relatively large aggregations of material, any one molecule is at any one instant within striking distance, so to speak, of but a small fraction of the other molecules. The rate of substance-transformation depends therefore not only upon the attractive *forces* within the small limits of distance within which this attraction is effective, but also upon the migrations of the molecules, which expedite the substance-transformation much in the same way that stirring hastens the solution of a solid substance in water.

Another important factor in substance-transformation is the internal agitation of the molecule, or the movements of the constituent atoms.

There may be three principal kinds of motion of the atoms. To fix our ideas we will consider the case of a two-atom molecule, such as the molecule of hydrogen or of hydrochloric acid. First, there may be the revolution of the two atoms about a common centre; secondly, a vibration of the atoms about some mean position; and thirdly, a movement of translation common to both atoms, which is the translation of the molecule.

In taking account of the motions of the atoms and molecules, we see why a sensible interval of *time* is necessary for the completion of the chemical transformation of an appreciable quantity of substance. On Dalton's theory of chemical action it is not at first sight easy to understand why such chemical changes as those in an intimate mixture of different gases should not proceed with absolute suddenness. When the conditions necessary for transformation are attained, why are not *all* the molecules simultaneously transformed?

The answer is that the conditions of individual molecules in such a mixture are, at any moment, widely different. Some are travelling slowly, some fast. In some molecules the atoms are revolving more rapidly round the common centre than in others, and consequently we should expect the distance between the atoms in some molecules to be greater than the distance between the atoms in other molecules, of the same composition. The mutual attraction of the atoms would therefore be less, and they would be therefore more ready to part company. Similarly the amplitude of the vibrations of the atoms will be greater in some molecules than in others. Probably the amplitude of the vibrations will be greatest in those molecules which are travelling fastest, in which the velocity of revolution will also, probably, be greatest. The molecules which travel fastest, undergo the most frequent encounters, in other words, are most frequently brought within such distance of other molecules, that atomic attraction between them is effective. This last consideration applies more particularly to gases, from which our knowledge of molecular motion is almost entirely derived. It appears then that the molecules which move fastest are, in respect of all the three conditions of motion we have specified, most liable to chemical change.

The kinetic theory of gases shows that, at any one time, a *small proportion* of the total number of molecules in a gas, perhaps also in solids and liquids, are moving with relatively great velocities. A small proportion of the molecules will therefore be in a condition to undergo chemical change before the main body of their fellows has attained the same state.

If the chemical change be accompanied by evolution of heat, the combination of a few molecules will tend to raise the temperature and quicken the travelling motion and rotation of other molecules. In such cases the reaction, started among a few molecules, is propagated throughout the whole body of the molecules, and may continue without any further addition of external energy. This is the case more particularly in what are called explosive mixtures, such as hydrogen and oxygen, or hydrogen and chlorine, in which the rate of substance-transformation may be very great. When the intense, but merely local, heating effect of the electric spark, fires an explosive mixture of hydrogen and oxygen, the heat given out by the reaction in the immediate neighbourhood of the spark, is sufficient to set the reaction going at a rapid rate throughout the whole body of molecules. Similarly, a mixture of about equal volumes of hydrogen and chlorine, combines with explosive rapidity under the influence of bright sunlight. If, however, the gases are only exposed to diffuse daylight, the combination is gradual.

Again, if an explosive mixture of gases be diluted with a large excess of one or other component, or with a large excess of a third gas, which does not react with them, then the effect of the initial action is damped down. If the excess of indifferent gas be sufficiently great, the spark, or the beam of sunlight, may even prove insufficient to initiate a perceptible action.

We see, therefore, that the rate at which a substance-transformation proceeds, is conditioned by circumstances such as dilution, as well as by the magnitude of the atomic attractions.

The *total amount* of energy developed in a particular reaction, during the transformation of a given amount of substance, is constant; but the rate of development of energy is dependent upon circumstances. Thus, when combustible materials are slowly oxidised, as food is oxidised in the body, the *rate of work* must be small compared to what one would get if all the atoms had the opportunity of bringing their attractive forces simultaneously into play.

We get our nearest representation of the magnitude of atomic forces, of what we may call their *proper* rate of working, in the detonation of modern explosives.

When a chemical change is accompanied by absorption of heat (*e.g.* the splitting up of carbonic acid in presence of the green colouring matter of plants), the action of the first few reacting molecules does not propagate the change throughout the whole body of molecules. More energy must be added from outside, if the substance-transformation is to proceed. In the case cited, the splitting up of carbonic acid ceases, when the sun's radiation is withdrawn. Reactions of this class (called *endothermic*, because they are accompanied by absorption of heat) are not a direct source of available or kinetic energy. On the contrary, they use up active energy, and store up latent or potential energy.

The great store of energy which we possess in the capacity of fuel to unite chemically with oxygen, is due to the chemical mechanism which enables plants to store up the energy of solar radiation, which, in the tropics, attains about two-horse power per square yard. Plants are the great "accumulators" of energy, and this *endothermic* reaction, the splitting up of carbonic acid, is more important to the economy of that part of nature with which mankind is brought in contact than any other chemical change.

The case of a chemical change, accompanied by *absorption of heat*, appears at first to present a difficulty in the way of accepting any simple doctrine of a chemical attraction between atoms. If work has to be done upon two atoms to cause them to combine, it would at first sight appear that there must be repulsion, not attraction, between them.   If we were dealing with reactions between one-atom molecules of elements, this argument would perhaps be admissible; but the argument is of no effect in the case of reactions between complex molecules.

Acetylene is formed with absorption of heat, yet there is no reason to suppose that the carbon and hydrogen atoms are combined chemically against a force of repulsion subsisting between them.   The heat-absorption only shows that more energy is required to tear apart the carbon atoms in solid carbon and the hydrogen atoms of the hydrogen molecule, than is given out by the joining up of the carbon to the hydrogen atoms.   If there were repulsion between the carbon and hydrogen atoms, in place of chemical attraction, it is difficult to escape from the conclusion that the molecule of acetylene would cease to exist unless external energy were constantly supplied to it.

Similarly, when a compound decomposes with evolution of heat, as, for instance, is the case with chlorine monoxide gas (Cl_2O), this is not because the chlorine and oxygen atoms are held together by any external coercive force, but because the energy developed by the formation of mole cules of free chlorine and free oxygen is greater than the energy needed to separate the oxygen atoms from chlorine atoms.   In the case of endothermic compounds, such as chlorine monoxide, a certain amount of energy is needed to *start* the decomposition.   The new combination into which the atoms fall, gives out sufficient heat to continue the decomposition, through the operation of the increased velocities of translation, revolution, and vibration of the atoms.

It seems that the power of chemical attraction in atoms is something of the nature of a *charge*, which can be wholly or partially saturated.   It can hardly be doubted that the chemical attraction of atoms is intimately connected with electrical attraction.   Chemical force may in its ultimate nature be nothing but electrical force.   Electrical science affords the most hopeful means of accounting for the mode of transmission of chemical force from atom to atom. During the electrolysis of a solution of barium chloride ($BaCl_2$) *equivalent* weights of barium and chlorine are brought to the two electrodes, carrying *equal* charges of electricity, positive electricity in the one case, negative in the other. Similarly when potassium chloride (KCl) is electrolysed equivalent weights of potassium and chlorine carry equal quantities of electricity. The atom of chlorine carries a charge of electricity equal to the charge of the atom of potassium, and *one half* the charge of the barium atom. This shows the ultimate connection between *valency* and electrical phenomena. Those atoms which can hold in chemical union, two, three, or four atoms of hydrogen, chlorine or potassium, are capable of carrying two, three, or four unit charges of electricity. These phenomena are shown by such liquids as molten salts as well as by solutions.

These atomic charges of electricity are sufficiently great to account for the observed forces of chemical attraction (*vide* Lodge, "Modern Views of Electricity"), and are immensely greater than the force of gravitation calculated for atomic distances. Some measurements which have been made of atomic charges in gases are of the same order of magnitude as the atomic charge in solutions and liquids.

It seems as if the decomposition of a molecule into two parts, such, *e.g.*, as two atoms, or an atom and a radicle, is necessarily accompanied by the appearance of equal and opposite electrical charges.* It does not appear, however,

---

* The following passage from Ostwald is the best statement I am acquainted with of the facts which indicate that dissociation into electrically charged atoms is not due to the action of the electric current applied to the solution.

"Imagine two insulated vessels A and B filled with a solution of potassium chloride, and electrically connected by means of a syphon. Let a negatively charged body be brought near A ; remove the syphon, and, lastly, the charged body. Then, as is well known, A remains positively electrified, and B negatively electrified. Now, according to

that the power of atoms and radicles to combine so as to form molecules is dependent upon their possessing the *electrolytic* charges.   In the case of the electrolysis of a solution of sulphuric acid, the hydrogen atoms give up these charges to the electrode, and *then* are able to combine chemically, forming hydrogen gas.

Hertz, in his experiments upon electric waves, determined the wave-length of the disturbances emitted by electric oscillators of known dimensions.   From his results can be calculated the dimensions of an oscillator which would emit electric disturbances of the wave-length of light. These dimensions are of the same order as the size of a molecule.   It has been suggested that light waves are due to electric surgings in the atoms.

Lord Kelvin, in a discussion of phenomena occurring in highly rarefied gases, remarks that " it seems certainly true that without the molecules there could be no current, and *without the molecule electricity has no meaning."*

Chemical and electrical force are almost certainly connected in an intimate manner.

Faraday's law, the electricity in electrolytes can only move simultaneously with the ions.   Consequently, if an excess of positive electricity is present in A there must also be an excess of free positive ions, *i.e.* of potassium atoms, by the electricity of which the charge is determined.   If the electricity is conducted away, the potassium assumes the ordinary form, and acting on the water of the solution develops hydrogen, which can be collected in suitable apparatus and tested.

" Similar considerations hold good for the chlorine in vessel B.

" It is, consequently, not only conceivable that the ions in an electrolytic solution move about with electric charges, otherwise quite free, but solutions may be prepared which contain an excess of any ion we choose, *e.g.* an excess of potassium.   The assumption that electroytes contain free ions is not only possible, but necessary."

CHAPTER VIII.

CHEMICAL MANUFACTURE.

Sect. I.—The Fuel of Blast Furnaces.

In the first essay on Chemistry we traced the progress of our knowledge of the element carbon, from the time when it was recognised as a reducing principle by means of which ores could be made to yield the metals, to the time when Lavoisier showed that in this process of reduction the element carbon abstracts the oxygen from the ore, forming carbonic acid.

It was pointed out, also, that in such *reducing* processes the carbon plays a twofold part, for while one portion unites with the oxygen of the air, and in this process generates heat, the high temperature thus brought about enables the remainder of the carbon to abstract the oxygen from the ore, forming more carbonic acid, and leaving the metal. It is merely a matter of arrangement and of disposition of the parts of apparatus whether one portion of carbon is used as the heating, and a separate portion as the reducing agent, or whether the two processes are carried on simultaneously in the same parcel of the material. The former condition obtains, for example, when a powdered metallic oxide mixed with finely divided charcoal is placed in a glass or porcelain tube, which is heated in a charcoal furnace. The second condition would obtain when the common hæmatite iron ore (an oxide) was mixed with charcoal in the old-fashioned iron furnace of Spain, and the charcoal having been set

I

alight, and the fire having been urged by a blast of air, the iron was reduced to the metallic state by the action of the carbon.

In any process of iron smelting in which an oxide of iron is reduced by some form of carbon the initial and the final states may be represented thus—

Oxide of iron + carbon = metallic iron + carbonic acid.

The symbol = must here be interpreted " may be caused to yield." But, as our acquaintance with a chemical process becomes more intimate we generally find that a knowledge of the initial and final states of matter is not sufficient either for a proper understanding of what goes on, or, in many cases of manufacturing processes, for regulating the reactions so as to obtain a good " yield" of the valuable product. For this it is often necessary to learn the steps between the initial and final stages, even if the intermediate bodies formed have but a transitory existence.

This is the case in the reduction of iron ore in the blast furnace. The modern improvements are based upon the conditions of formation and decomposition of the lower oxide of carbon, carbon monoxide gas, which is formed either in the incomplete combustion of carbon, or when highly heated carbon acts upon carbonic acid. Carbon monoxide gas readily takes fire in presence of oxygen, burning with formation of carbonic acid. In so doing, the carbon takes up a second dose of oxygen, being combined in carbonic acid with just twice as much oxygen as in the lower oxide. Carbon monoxide also acts as a reducing agent, the gas at a high temperature extracting the oxygen from metallic ores.

If, in any metallurgical process a portion of the coal, coke, or charcoal is only burnt to carbon monoxide, then, in respect of that portion, we only get a part of the heating effect of the fuel, and half the reducing effect of the reducing agent. True, the hot carbon monoxide issuing from the apparatus burns in the air, forming the final product

BLAST FURNACES AT CLARENCE IRONWORKS.

carbonic acid, and producing more heat; but we want that heating effect in the apparatus, not outside it, and we wish, if it be possible, that the extra dose of oxygen required for complete combustion of the monoxide should come from the ore and not from the atmosphere.

In the blast furnace the materials put in at the top are oxide of iron, coke, and limestone.

The lime of the limestone, combining with siliceous matter mixed with the ore, removes these impurities in the form of a light fusible slag, which floats upon the top of the liquid iron collected at the bottom of the furnace.

The air is supplied in a blast forced through tubes called twyeres, placed near the bottom of the furnace.

The furnace is always kept full, or nearly full, and we may consider the conditions at any part of the furnace as remaining constant, as, when once lighted, the furnace is always kept in operation till it is "blown. out," after, perhaps, twelve or fourteen years' work.

The furnaces used at the beginning of the century were about forty feet high. Those used now are eighty feet high. It was found that by giving more time to the ascending current of carbon monoxide, a greater portion of ore was reduced. A great deal of the gas in the smaller furnace escaped without having an opportunity of becoming fully oxidised.

A still more important improvement was the introduction of the *hot blast*. It appears paradoxical that one ton of coke burnt to heat the air before entering the furnace, should save three tons of coke in the furnace itself; yet so it is. The heating (with hot blast) being partly done from outside, it is not necessary to put so much coke into the furnace; consequently, there is room for more of the ore. The power of the ore and limestone to intercept heat is double that of the coke which has been replaced, and there is a greater surface of the oxide to act on the diminished and slower current of carbon monoxide. Hence, economy in the production of pig-iron was effected by the use of the hot blast, and by raising the height of the furnace to eighty feet.

Some furnaces were built of more than one hundred feet, but it was found that no further economy was effected. The present approved dimensions of furnace, and temperature of blast, appear to give a better yield of metal than if the furnace be larger and the blast hotter, and a better yield than if the furnace be smaller and the blast colder.

The reason is as follows: We have together in the furnace, carbon monoxide and iron ore, carbon dioxide and coke. If the heat contained in the carbon monoxide is more completely intercepted before passing out of the furnace, then, as we have seen, more of the ore is reduced, and more of the monoxide is burnt to dioxide. But, on the other hand, if the temperature of the carbon monoxide be still further increased by the use of a still hotter blast, and if the heat of the up-flowing carbon monoxide be yet more completely intercepted before the gas issues at the mouth of the furnace, then we find that the temperature of the other materials is raised so far that the incandescent coke reduces the carbonic acid to carbon monoxide as fast as the ore oxodises the monoxide to carbon dioxide. Consequently we have two opposite reactions which balance one another under certain conditions.

When the temperature rises so much that the reduction of carbon dioxide is more rapid than the oxidation of carbon monoxide, it is obvious that we have passed the point where the furnace works economically. It appears, therefore, as the result of practical experience, that we must always be content to only half burn in the blast furnace a large proportion of the fuel, sending out a *mixture* of carbon monoxide and of carbon dioxide from the mouth of the furnace.

One of the first great improvements in blast-furnace practice was that of heating the blast. A more recent improvement has been to utilise outside the furnace the burning of that proportion of the carbon monoxide, the presence of which in the escaping gases is, as we have shown, a necessity. In the modern furnaces the mixed carbon monoxide and carbon dioxide are drawn off near the top of the furnace, instead of being allowed to come in contact with the air. The hot gases are only admitted to the presence of oxygen when they have been brought by pipes to the furnace where the air for the blast is heated. The combustion of the carbon monoxide not only heats the air for the blast, but furnishes power to drive the blast engine.

The arrangements for utilising the formerly "waste" gases of the blast furnace have effected a saving of more than half the total heating power of the coke, a saving equal, in Great Britain alone, to about four million tons of coal per annum.

SECT. II.—EXPLOSIVES.

The general phenomena of explosion are well known : noise, shock, resistance offered and overcome, and, in general, some work of destruction.

Various materials, solid and liquid, are employed for the purpose of producing an explosion. Every such material is termed an *explosive*, although sometimes, as in the case of gunpowder, the material is a mixture of several substances. Coal gas and fire damp are not called explosives, although on mixing with air they are capable of exploding. An explosive proper contains its own supply of oxygen.

Explosives are of two principal kinds, of which gun-powder and gun-cotton are typical examples. Gunpowder is a mixture of three different chemical substances, whereas gun-cotton (and similarly other nitro-explosives, as, *e.g.*, nitro-glycerin) is a single chemical substance, composed, however, of several elements, one of these being oxygen.

In gunpowder we have very intimately mixed together, two substances, charcoal and sulphur, which are capable of combining chemically with the oxygen contained in the third substance, nitre. All that is necessary to bring about this chemical change is, firstly, that the particles of the various substances shall be brought very close together, which is effected by the careful incorporation of the ingredients ; and, secondly, that the temperature should be high.   At a high temperature gunpowder takes fire, or, if the conditions be suitable, explodes.   In the combustion of gunpowder the carbon burns to carbonic acid (a gas), the sulphur burns to sulphurous acid (a gas), and other gaseous products are also formed in considerable quantity.

The heat developed by the chemical reaction raises the temperature of the gases, which causes them to expand rapidly and occupy a large volume.   If the burning of the gunpowder takes place in a confined space, the force of expansion of the gas is resisted by the walls of the enclosure, and if the latter are not of sufficient strength, they are overthrown.   If, however, the enclosure be sufficiently strong and furnished with no outlet (as in certain experiments where very small charges are fired inside strong bombs), the gases formed are kept compressed in a small volume, and no disruption takes place.   In this case we have an explosion, without its usual striking accompaniments.

When a cartridge is fired in a gun-barrel, the gases, whose force of expansion would be sufficient to burst the steel of the barrel, find in one direction no great resistance, and expanding rapidly in this direction, propel before them the bullet, which thus leaves the muzzle of the gun with a high velocity.

Explosives are compounds or mixtures, which contain in themselves elements capable of taking up a new molecular arrangement forming fresh compounds, one or more of the new compounds being gases, the formation of these compounds being accompanied by an evolution of heat. The formation of gases, and the development of heat in the reaction, are essential to the production of an explosion.

The condition generally necessary to produce the explosion of an explosive body appears to be the rapid vibration of its particles.   Such vibration may be generated by heat, by shock, or by friction. A sudden blow will serve to detonate dynamite, although the heating effect of the blow may be very slight. The liability to the occurrence of explosion under such circumstances depends not only on the *force* of the blow, but also on the nature of the striking bodies. Thus, a force of blow which would cause explosion if the blow were one of steel against steel might be harmless if produced by wood against wood. In this case the explosion is probably determined, not by the amount of heat produced, but by the rapidity of the vibration.

When an explosion occurs, the chemical atoms are *shaken out* of one combination to fall into another. This process takes place much more readily when the explosive is warm.

Although explosions are often produced independently of heating effects, yet at a sufficiently high temperature all explosives will detonate.

In the study of phenomena connected with explosives we often meet with occurrences at variance with the old dog-Latin dogma, *causa æqual effectum*. The pulling of the trigger, followed by the rush of the bullet from the gun, is a familiar example; the work done in pulling bears no proportion to the energy developed by the explosion.

Another case in point is afforded by the manner in which explosions are sometimes caused in the incorporating mills, in which the component materials of gunpowder are mixed together. The presence of a small hard body, such as a nail, or even a hard piece of grit, may cause sufficient local heating to start an explosion of the whole mass. Both shocks and local heating have to be carefully guarded against in gunpowder factories.

Charcoal possesses the property of condensing air in its pores, which sometimes leads to local heating and induces spontaneous combustion.

In grinding the sulphur there is another source of danger. Sulphur is a highly electric body, and in the process of grinding a large amount of electricity will often accumulate, sometimes giving rise to sparks, the passing of which may produce serious consequences. The danger from this source is, however, to a great extent overcome by connecting the sulphur mills to earth by means of wires, and thus continually drawing off the charge of electricity produced by friction in the grinding process.

In the pressing of the gunpowder, hydraulic machines furnished with ebonite plates are frequently employed. Ebonite is a convenient material for the purpose, being tough, elastic, smooth, and sufficiently hard. Unfortunately, ebonite is a highly electrical material, and the upper and lower plate, with the cake of powder between them, practically form an electric pile. A passing thunderstorm may induce a discharge of sparks from the ebonite, igniting the gunpowder and producing, as has happened in several cases, fatal accidents.

In spite of all precautions, explosions are liable to occur in the mixing of the materials for gunpowder, and it is well to provide, as far as possible, for the safety of the *employés* and of the building. A good protector for the workmen is a curtain of ships' hawsers, which offers the kind of resistance which is most effectual in the case of an explosive outburst of gas.

By having a light roof, secured only by one or two wooden pins, an outlet is obtained for the gases produced in a factory explosion. The roof is simply lifted off, and the outlet thus given prevents the pressure inside from becoming sufficiently great to damage seriously the main portions of the building.

Nitro bodies explode more readily under shock, and at a lower temperature than gunpowder. Nevertheless, in the hands of properly trained workmen, the manufacture of nitro-glycerin and dynamite is accompanied by fewer casualties than that of gunpowder. Gun-cotton is prepared by the action of strong nitric acid and sulphuric acid upon cotton-wool. Most of the processes are carried out in presence of a large excess of water, though this is, of course, not the case during compression, in which operation great care has to be exercised. Nitro-glycerin, prepared by acting upon glycerin with a mixture of nitric and sulphuric acids, is liable to explode both by heat and by shock.

Dynamite is produced by absorbing three parts of nitro-glycerin by one part of kiesulguhr, a finely divided siliceous earth capable of absorbing a large quantity of liquid without becoming pasty. Dynamite only explodes when subjected to special treatment, being unaffected by moderate heat or by an ordinary blow, but detonating under the sharp shock given by a percussion fuse of fulminating mercury. The kiesulguhr plays no part in the actual explosion, so that dynamite, as an explosive, must be classed along with gun-cotton and the other *compounds*, rather than with explosive *mixtures* such as gun-powder.

A special class of explosives is required for filling percussion caps and detonators. Fulminate of mercury is the most important of these highly dangerous substances, the manufacture of which is conducted with the most elaborate precautions, not only against shock, but against the smallest amount of friction.

The protection of the factories against lightning is a problem of considerable difficulty. According to Mr. O. Guttmann, to whose writings I am indebted, a system similar to that of Professor Lodge's "network" protector has been extensively and successfully used by the Austrian military authorities. The system is similar to that by which electrometers are shielded from electrification by means of a wire cage, the building being covered by a network of galvanised iron wire. This material is, of course, much cheaper than copper, and its smaller electric conductivity does not appear to be a serious drawback in the case of electric discharges of such high potential as that of lightning.

Sect. III.—Bye-Products and Waste-Products.

The well-known Leblanc process for making soda came into use about one hundred years ago. Before the introduction of this process the alkali required in various manufactures was nearly all obtained from the ashes of plants. Potash, the alkali obtained from land plants, was more plentiful than soda, which is obtained from the ash of sea plants. The sodium of the soda yielded by sea-plants is contained in the sea in the form of common salt.

By Leblanc's process the soda is prepared directly from common salt, thus avoiding the tedious and uncertain collection of sea-weed. The first impetus which the new method received was in 1793, when France found herself deprived of alkali, which used to come chiefly from Russia and America. How serious was the inconvenience thus caused may be readily imagined, since the manufacture of soap and of glass is dependent on a supply of this material.

Leblanc's process for preparing soda directly from salt relieved France from an industrial difficulty. It also changed the relative importance of the two alkalies, making soda much cheaper and more plentiful than potash. Leblanc himself died in a French workhouse, but manufacturers on this side of the Channel were more fortunate, and many of them accumulated wealth as rapidly as the iron-masters did in later times. The demand for soda was unlimited, and the sole object of the manufacturer was to produce as much of the article as possible.

The materials for the manufacture are the following, viz.: first, salt and sulphuric acid, which react on one another, producing "salt cake" (sulphate of soda) and hydrochloric acid gas; secondly, limestone and coal, which, being heated with the sulphate of soda, produce the required alkali carbonate of soda, more commonly known simply as "soda." In this second reaction the final products, besides soda, are sulphide of calcium and carbonic acid gas.

Leaving out of account this last substance, the product of ordinary combustion, we see that in the manufacture of soda the waste products are, or were, two, viz. hydrochloric acid gas and sulphide of calcium, or "alkalimakers' waste." The former substance was allowed to escape into the atmosphere, killing all vegetation for miles around, whilst the latter accumulated in vast heaps in the neighbourhood of the works, constantly undergoing decomposition through the action of air and moisture, and poisoning the atmosphere with fumes of sulphuretted hydrogen, besides polluting the streams with the poisonous drainage of the decomposing mass. The alkali-maker cared for none of these things, and the Government found it necessary to legislate in order to protect the health and property of the manufacturer's neighbours. The manufacturers were compelled to condense and retain the fumes of hydrochloric acid, though the decomposition from the "waste" continued to be a nuisance to the neighbourhood.

The alkali-maker being obliged to go to the expense of collecting the hydrochloric acid, set to work to compensate himself for the outlay involved, by utilising the formerly waste product, which now became a bye-product of the manufacture.

More recently a process has been introduced for recovering the valuable sulphur from the sulphide of calcium. Employing the same materials as formerly (salt, sulphuric acid, limestone, and coal), the principal products of the modern manufacture are three instead of one as formerly— that is to say, soda, hydrochloric acid, and sulphur; instead of soda only.

From the hydrochloric acid chlorine is now made, and the chlorine gas, passed over dry slacked lime, forms the bleaching powder which is used in enormous quantities for whitening cotton and other goods. Large quantities are also used for disinfecting purposes. At first one of the chief items of expense in the manufacture of bleaching powder was the consumption of manganese dioxide, employed to set free chlorine gas from the hydrochloric acid. This is now "regenerated" by Weldon's process, in which the chloride of manganese formed in the reaction is acted upon by air and steam in the presence of lime, the final "waste" material being calcium chloride.

Similarly, in the process for the recovery of sulphur from the alkali-makers' waste, the same substance, calcium chloride, is the final waste product. The sulphur-recovery process is conducted in two stages: first, the partial oxidation of the sulphide of calcium by air and steam, and secondly, the decomposition of the oxidised product by hydrochloric acid, when sulphur separates out and calcium chloride is formed.

Calcium chloride is a material for which there are but few applications, and it is still practically "waste." The loss of calcium is not of importance, since abundance of the useful compounds of this element can always be obtained from natural sources, such as limestone. But the chlorine is valuable, and many efforts have been made to recover this element from calcium chloride. It is quite possible to set the gas free, but hitherto a sufficiently cheap and simple process has not been brought into use, although much attention has been devoted to the problem.* Calcium chloride is the final waste product in a large number of chemical processes.

The soda made by the Leblanc manufacturers would now be produced at a loss, were it not for the utilisation of bye-products. This is due to the introduction of another method of manufacture, known as the ammonia-soda process, in

* A new method for obtaining the whole of the chlorine from common salt is based upon the oxidation of the hydrochloric acid by nitric acid, instead of by manganese dioxide.

which a purer product, commanding a higher price, is simply and quickly obtained.

In this process carbonic acid gas is passed through water which contains, dissolved, both common salt and ammonia. Soda, in the form of the bicarbonate, is deposited from the solution in crystals, the carbonic acid uniting with the base soda, and the ammonia taking the chlorine, forming ammonium chloride, which remains in solution. The conditions under which the soda is produced in this method of manufacture are very favourable to the formation of a pure product. Both by the solution of the common salt and by the subsequent crystallisation of the carbonate of soda, impurities are eliminated. In the Leblanc process, on the other hand, impurities accumulate, and their subsequent removal involves time and expense. The second stage of the Leblanc process, as we have already stated, consists in the decomposition of sulphate of soda by means of limestone and coal. All these solid substances are heated together on the floor of a furnace. After cooling, the mass is treated with water, which dissolves out the carbonate of soda and leaves behind the calcium sulphide and any unburnt coal. But the solution contains a good many other substances besides carbonate of soda, such as sodium sulphide and sodium thiocyanate, besides large quantities of caustic soda.

A full account of the various inventions for the utilisation of bye-products would fill a volume, and does largely contribute to fill many volumes of technical literature. In the blast furnace, the constant endeavour has been to utilise the energy of the half-burnt carbon so as to diminish the consumption of coal. In the Leblanc process the economy has been effected by the preparation of useful substances from useless materials. The iron-makers, like the alkali-makers, had hard times to meet, and they tided over the hardest time by reason of the improvements which care and foresight led them to introduce in the manufacture. The alkali-maker only turned his attention to economy of material when legislation compelled him to do so ; but in his case, and in that of the more provident producer of iron (and, we might add, of manufacturers in all branches of industrial chemistry), prolonged prosperity has only been attained by minimising wastefulness as much as the wasteful tendency of all natural change permits.

CHAPTER IX.

Sect. I.—Animal Heat.

Lavoisier first proved to the world the nature of the chemical changes which occur in burning, and showed that in respiration the same chemical process (oxidation) is going on.

Every chemical change is accompanied by a heat change. Lavoisier not only recognised this fact, but he appreciated its importance more than his immediate successors. Accordingly he set himself to investigate *quantitatively* the heat changes which occur in the phenomena of burning and of respiration, both being, from his point of view, phenomena of oxidation.

How well Lavoisier understood that without measurement there is no science is shown by the fact that he used the *calorimeter*, or instrument for measuring quantities of heat, with the same diligence as he showed in the use of the balance.

Calorimetry, that branch of science which deals with the measurement of quantities of heat, owes its origin to the joint work of Lavoisier and Laplace, who were the inventors of the ice-calorimeter. The principle of their method was to measure the quantity of heat by the weight of the ice melted. Thus, in order to determine the quantity of heat given out from one pound of iron in cooling from the temperature of boiling water to that of melting ice, the hot iron was placed in the central chamber of a box furnished with a lid and having two annular chambers outside the

central chamber. The outer annular chamber or compartment was filled with ice to protect the next compartment,
also filled with ice, from the warming effects of the external
air. All the heat received by the inner ice compartment
comes, therefore, from the body placed in the central
chamber. The weight of ice melted in the inner compartment measures the quantity of heat received, the water
formed by the melting of the ice being run off from the
bottom of the chamber, which is funnel-shaped and provided with a stop-cock for the purpose. The above arrangements served for the comparison of the specific
capacities for heat of different materials, as shown by the
heat given out by them in cooling through a certain range
of temperature.

In the experiments on the heat given out in burning and
in respiration, the arrangements were slightly different. We
give Lavoisier's description, retaining the old term, *caloric*,
which may be taken to mean "heat considered as a measur-
able quantity." Lavoisier writes: " To determine the
quantity of caloric disengaged during combustion and
during animal respiration, the combustible bodies are
burnt, or the animals are made to breathe, in the interior
cavity, and the water produced is carefully collected.
Guinea-pigs, which resist the effects of cold extremely well,
are well adapted for this experiment. As the continual
renewal of air is absolutely necessary in such experiments,
we blow fresh air into the interior cavity of the calorimeter
by means of a pipe destined for that purpose, and allow it
to escape through another pipe of the same kind ; and that
the warmth of the air may not produce errors in the result
of the experiments, the tube which conveys air into the
machine is made to pass through pounded ice, that it may
be reduced to 32° F. before it arrives at the calorimeter.
The air which escapes must likewise be made to pass
through a tube surrounded by ice included in the interior
cavity of the machine, and the water which is there produced must make a part of what is collected, because the

caloric disengaged from the air is part of the product of the experiment."

By means of this apparatus, or "machine," as he calls it, Lavoisier compared the quantities of heat evolved during the burning of known weights of carbon and hydrogen, and of known weights of animal and vegetable substances formed of carbon and hydrogen, such as wax in a wax taper, and olive oil burnt in a little lamp.

He also determined "the quantity of caloric disengaged during respiration," a research connected with that on the loss of weight during respiration, in which his colleague Seguin was the *corpus vile* in place of the guinea-pig, which had been found so "well adapted" for calorimetry.

Seguin was sewn up in a varnished air-tight silk bag, the edges of which were accurately cemented round his mouth, leaving only a slit for breathing. He was weighed in a delicate balance from time to time.

Since the days of Lavoisier thermo-chemistry, or the study of the heat-changes which accompany changes of chemical composition, has made considerable progress, in spite of the many practical difficulties which surround the subject. One of the most important of the laws which have been experimentally established is that the initial and final stages of a chemical reaction alone determine the amount of the heat change.

For instance, the conversion of a given weight of carbon to carbonic acid is accompanied by the evolution of a quantity of heat which is the same whether the carbon be burnt rapidly in oxygen, or whether in a slow and roundabout series of chemical changes the carbon is successively a constituent of a number of vegetable and animal substances before finally attaining the form of carbonic acid in the expired breath of animals.

This law, which has been experimentally proved to hold in a variety of cases, enables us to calculate the heat-giving power of foods without having to ascertain the various changes and modifications which the food

K

undergoes in the animal body. We know the final products, and that is sufficient.

Thus, the sugar taken into the body is sooner or later completely converted into carbonic acid and water; in other words, it is completely burnt. The heat given to the body by one ounce of sugar is therefore the same as the heat evolved by burning an ounce of sugar. This quantity can be experimentally determined by the use of Lavoisier's calorimeter, or one of the modern improvements upon his original apparatus. We are thus enabled readily to compare the heat-giving power of different foods and food stuffs.

Every chemical change is accompanied by a heat change, but it must not therefore be supposed that every chemical reaction is accompanied by an evolution of heat. There may equally well be a disappearance of heat. It is the former class of reactions, those in which heat is evolved, which are often accompanied by striking phenomena such as evolution of light, etc. Hence the popular idea that a chemical change is necessarily accompanied by an evolution of heat.

By the second law of thermo-chemistry the initial and final stages only determine the heat reaction. Take the case of the carbonic acid of the atmosphere, and the various changes the carbon undergoes after its assimilation by the green portion of a plant and during its subsequent changes in the body of some animal which has fed upon the plant. The final stage is that the carbon is restored to the air as carbonic acid. The initial and final stages are identical and the total heat reaction is *nil.* While the carbon has been in the animal body it has been gradually oxidised up to carbonic acid, and during the whole of this time heat is being evolved. During the process of converting the carbon of carbonic acid into the state of chemical combination in which it is found in the plant heat is absorbed.

How is it, then, that the frigorific effects of plant growth are so much less patent and obvious than the heating effects of animal life? It is mainly due to the circumstance that, in order to bring about chemical changes in which there is an absorption of heat, some external agent must act, and must keep on acting, in order that the chemical reaction may proceed, and in the case of the decomposition of carbonic acid in the presence of the green colouring matter of plants this external agent is the sun's radiation. Hence, the cooling effect which results from the feeding of plants upon their atmospheric food is to a great extent masked by the fact that the cooling action only takes place in presence of a potent heating agent—the sun's rays. Still, the day temperature in a forest is much lower than that of a sandy desert in the same latitude. This gives some idea of the cooling action of plant life, which contrasts so sharply with the warmth of living animals.*

Let us consider what chemical changes, and heat changes, take place after the death of a plant. Decay and decomposition, which are mostly processes of oxidation, are accompanied by heating effects. Many such cases can be called to mind, such as the heating of hay in the stack.

On the other hand, the surest evidence of death in animals is the loss of bodily warmth. Thus, the opposite function of plants and animals, with regard to heat, are more sharply contrasted in the phenomena of death than of life. The contrast must not be pushed too far, since

---

* Physical action co-operates with chemical causes to emphasise the cooling effects of vegetation. The evaporation of moisture in presence of the sun's rays is in sharp contrast to the *condensation* of the breath of animals which is accompanied by a heating effect.

The dependence of the cooling effects of vegetation upon *growth* seems to me to be very noticeable. A field of mowing grass not yet fit for the scythe is a deliciously cool place; a hayfield, on the other hand, is a very hot place, when the hay has been made.

On a blazing June day, under the young leaves of lime, or beech, or chestnut, one does not feel the heat as on a sunny August day, when the sun's rays seem to pierce the moribund leaves; and one misses the cool draught of air which in early summer draws along under the trees.

when decay sets in there is a heating effect with animals also.

Warmth and power of locomotion are marked and obvious characteristics of living animals distinguishing them from plants. The power of locomotion which animals possess depends upon their warmth. The ordinary movements of an animal are made against some kind of resistance, and when motion is so accomplished physical work is performed. Physical work is measured numerically by multiplying the amount of the force by the distance moved against its resistance. All bodily work may be expressed in this way. The systematic and continuous use of the rational and thinking faculties is often called work, but for present purposes we exclude brain labour from the scope of the term "work," which we shall employ in the physical sense only. Combinations of mechanism (of pulley, screw, lever, etc.) cannot do work, they only vary the proportion of the two factors of work, distance and force ; but engines, like animals, actually do work, for an engine is provided with a source of energy as well as with a train of mechanism. A steam engine, as far as chemical science and heat science are concerned, is closely analogous to an animal. From this point of view, it is an animal very much simplified. Energy is obtained in the steam engine and in animals by burning carbonaceous material. Both the engine and the animal can do work as long as chemical action furnishes them with energy. Without fuel or food both grow cold, and no work can be done.

By the study of the steam engine, Hirn proved that every foot-pound of work done entails the disappearance of a definite quantity of heat between the boiler and the condenser of the engine. The relation is about 1370 foot-pounds of work for each unit of heat, the unit of heat being that quantity which will raise one pound of water through one degree, centigrade.

Hirn's method was as follows : He determined the total amount of heat put into the boiler from the amount of the water supplied per diem and from the temperature of the steam. The amount of heat returned to the condenser was calculated from similar data.

The amount of mechanical work done was measured by a Watt's *indicator*. In this instrument the amount of work done is shown by the movement of a pencil point upon a sheet of paper. The position of the pencil point, which indicates the work done, is determined jointly by the pressure or *force* of the steam, and the *distance* through which the working piston has moved. The work registered by the indicator is partly the work done in overcoming the resistance of the parts of the engine, which is called internal work, and partly external work, such as raising a weight or driving a shaft.

Hirn found that, if he increased the external work of his engine, there was a greater loss of heat between the boiler and condenser, one unit quantity of heat disappearing for each 1370 foot-pounds of additional external work. A great variety of methods has given practically the same value for the mechanical equivalent of heat, and we may apply the value obtained to the solution of problems upon the relation of food and work in animals.

In the animal body there is "internal work" to be done, just as there is in the steam engine. In neither case can the whole of the heat furnished by the fuel or food be converted into external mechanical effect. This would require, in the case of the steam engine, that no heat at all should reach the condenser, and that the condenser should be at the absolute zero of temperature, a condition which it is impossible practically to attain. In the case of an animal, if all the heating power of its food were used for external work the temperature of the animal's body would sink to that of its inanimate surroundings, a condition obviously incompatible with life. As a matter of fact, about one-fifth of the energy which the food develops can be obtained in the form of external mechanical effect, the remainder being required for the internal work of the body. This is a much larger proportion of external effect than in the case of a steam engine, or other heat engine, where the *efficiency*, or proportion of the external work done to the mechanical equivalent of the heat supplied is more like one-twentieth.

The heat developed in the body by one pound of the carbohydrates (starch, etc.) has a mechanical equivalent of about 2860 foot-tons. One pound of oil or fat is in the same sense equivalent to 6450 foot-tons. About twenty per cent. of this is available for external work, so that we may say that if a man is to do three hundred and twelve foot-tons of work he may supply himself with the requisite store of energy by taking four ounces of fatty food in addition to the *maintenance* diet necessary to meet the daily waste of the body. Three hundred and twelve foot-tons is about the average amount of work done in a day by an English labourer, whose diet would probably comprise about four and a half ounces of fat.

It seems paradoxical that working should make a man hot rather than cold, seeing that heat is used up to produce mechanical effect. The analogy of the steam engine will enable us more readily to explain this apparent anomaly. When a locomotive is going sixty miles an hour, more work is being done than when it is going thirty miles an hour. More heat is converted into work when the engine is going fast than when it goes comparatively slowly; but no one would expect the heat in the engine as a whole to be less when the pace is great. The draught will be greater, and the consumption of coal greater; so much greater, in fact, that, although there will be more heat converted into mechanical effect by the working of the piston, yet the engine as a whole will be hotter.

It is so also in the case of the animal body. When bodily work is being done, the contracted (*i.e.* the working) muscle seizes upon a greater quantity of the oxygen of the arterial blood than in the case of the uncontracted muscle in its state of rest. The blood of the veins may still contain as much as seven and a half per cent. of oxygen when the muscles are at rest, but a working muscle may leave as little as one and a half per cent. of oxygen in the venous blood. This means that there must be a corresponding increase in the amount of carbonic acid given off by the lungs, which in fact may be as much as five times greater during work than during repose. Respiration is deeper and more rapid, more oxygen is inhaled, the combustion of the food goes on more quickly, and an increased supply of carbonaceous food is required to supply the fuel for combustion. Thus it is that the temperature of the body is maintained during work, although the work is done at the expense of food which otherwise would produce animal heat.

Sect. II.—Food-stuffs.

Our diet is so varied that it is not a very easy matter to find out what quantities of the necessary *food-stuffs* one eats.

The weight of food taken during the day may be most readily determined by using one of the spring balances employed for weighing letters and parcels. The plate or cup, with its charge of food, may be placed upon the disc of the letter-weighing balance, and after the experimenter has eaten all, or such part as satisfies him, the plate or cup is again placed on the disc, and the difference of the two weighings is entered in a note-book. With the spring balance the weight is read off at once on the dial, so that the weighing can be done sufficiently quickly not to let the food get cold.

Having thus conscientiously recorded the amount of food, solid and liquid, taken during a day, the next thing to do is to calculate out the amount of the various " food-stuffs " in the day's ration, and compare them with a standard ration such as may be found in a book on the chemistry of food. The daily ration required by a man depends upon the daily bodily waste, which, again, depends upon the weight and upon the amount of bodily work done.

The standard ration given below (see p. 137) is calculated for an eleven-stone man supposed to be taking moderate exercise. If hard bodily work is to be done, the amount of food containing starch and fat must be increased.

Though our foods are so varied, yet the different kinds of feeding materials or "food-stuffs" which they contain are few, or at least for practical purposes may be reckoned under a few heads. Thus—

FOOD-STUFFS are—

1. Albuminoids and other substances containing nitrogen.
2. Fat, starch, and sugar.
3. Mineral substances, chiefly common salt and phosphates.
4. Water.
5. Food adjuncts.

The principal functions of the above classes of food-stuffs are as follows :—

1. Albuminoids, etc., are oxidised by the air which is inhaled into the body, and go to form the muscle and flesh. The process of oxidation gives out heat, and hence these *flesh-forming* foods assist also to keep up the heat of the body.

2. Fat, starch, and sugar are oxidised in the body, thereby acting as *heat givers*, but they do not form muscle and flesh. In hard bodily work it is necessary to increase the supply of merely heat-giving food. The waste of muscle and flesh does not increase with increased bodily labour in nearly the same proportion as does the demand upon the heat of the body.

Fat has about $2\frac{1}{3}$ times the warming power of starch and sugar, there being a larger proportion of carbon and hydrogen to be oxidised. Hence fat is the principal food-stuff to be added to the diet to meet the demand of extra bodily work or the corresponding tax of a colder climate. Fat can be stored up in the body (as a layer under the skin), where it acts as a store of heat-giving material which can be drawn upon as required by the system.

3. Of the mineral substances taken into the body, the lime-salts and phosphates form the hard part of the bones.

4. Water constitutes about two-thirds by weight of the substances of the body. The water taken in as food is required both as a constituent of the body and also as a carrier of food in and through the system.

5. Food adjuncts are of importance more on account of their effects (whether stimulating or sedative) upon the nervous system and on account of their effect on the palate than for any actual power of nourishing or sustaining the fabric of the body. The most important are alcohol, the alkaloids contained in tea, coffee, cacao, and tobacco.

The daily ration of an adult under ordinary conditions, according to Professor Church, should contain the different food-stuffs in *about* the following quantities :—

DAILY RATION.

1.	Water	. . .	88·66	oz. avoirdupois
2.	Albuminoids .	.	4·25	,,
3.	Starch and sugar	.	11·40	,,
4.	Fat	. . .	3·77	,,
5.	Mineral food .	.	1·03	,,

109·11

The small quantity of food adjuncts is not included in the above table. The actual weight of food eaten will exceed the above total by about one ounce on account of fibrous material, either vegetable or animal, which is taken with the food, but which is not assimilated by the body. The food-stuffs 3 and 4 have the same function, viz. that of keeping up the heat of the body. Weight for weight, fat has about $2\frac{1}{3}$ times the heating power of starch ; and as starchy and fatty foods are to some extent interchangeable in diet, a convenient way of expressing the daily ration is to multiply the amount of fat by $2\frac{1}{3}$, which gives the quantity of starch equal to the fat in heat-giving value. In the above case the amount of fat is 3·77, which, multiplied by $2\frac{1}{3}$, gives 8·8, which, added to 11·40, the amount of starch and sugar, gives a total of 20·2 ounces of heat-giving food *reckoned as starch.* We may, therefore, write the daily ration, in ounces, thus—

		Heat-givers	
Water.	Flesh-formers.	(reckoned as starch).	Minerals.
88·66	4·25	20·2	1·03

Let us now return to the practical calculation of the quantities of the food-stuffs contained in a day's diet, the weight of which we have supposed to be recorded during meals by aid of the spring letter-balance.

The results of the weighings are set out in a tabular form, as follows : On the left, in the first vertical column, set down the names of the various foods taken ; bread, butter, milk, etc. In column two, set down opposite the names of the foods the weight of each which has been eaten during the day. In some work on the chemistry of foods find the percentage composition of the food eaten, and hence calculate the weight of each food-stuff in the several articles of food taken. Enter the water in vertical column number 3, the flesh-formers in column 4, the heat-formers in column 5, and the minerals in column 6.

The vertical columns are then added up, and we obtain at the foot of column 2 the total weight of food taken, and at the foot of the other columns the total weight of each food-stuff. It then remains to compare the results obtained with the standard ration given above, or with some other ration suited to the particular nature of the daily employment or the constitution of the person.

Of course, it is not only the quantity of the food-stuffs

which has to be taken into account in judging whether the diet is a suitable one or not; we must also take account of the proportion between the amounts of the different food-stuffs. The most important ratio, the value of which should, however, depend upon the amount of bodily work done, is the ratio of flesh-formers to heat-givers. If the fat be calculated according to its "starch equivalent" the ratio should be about $1:4\frac{3}{4}$. The ratio of flesh-formers to heat-givers is termed the nutrient ratio, or albumenoid ratio. A table of the nutrient ratios of different classes of food is a valuable guide in the determination of a diet. Such knowledge is absolutely necessary where a uniform diet has to be prescribed for persons who are not at liberty to select and vary their food according to their inclination, as, for instance, prisoners.

The most important flesh-formers are the lean of meat, and green vegetables. The starchy foods, as rice and potatoes, are, generally speaking, cheap and easily obtained. The fat of meat supplies the heat-givers in a more concentrated form, though not so cheaply.

The form in which the various food-stuffs may be taken must depend largely upon the mode of life. Thus, a labouring man who works in the fields may make his supper off cabbage and fat bacon, deriving the flesh-formers from the vegetable and the heat-givers from the animal food. For men of sedentary habits such a combination would be scarcely suitable. A lean chop with potatoes—a very usual luncheon for a business man—supplies the same food-stuffs in a lighter form. In this case the flesh-formers are divided from the animal food and the heat-givers from the vegetable. Needless to say, this is a more expensive meal than the first. Cheese is one of the cheapest forms of nitrogenous food, and may be taken as a substitute for lean meat, except in the case of people of sedentary occupation.

The effect of *cooking* is, in the first place, to soften the food, and thus render it easier to masticate. The heat may be applied as dry heat, as in roasting, or moist heat, as in steaming; but in either case the large quantity of water which is contained in all foods makes the various processes of cooking largely a matter of steaming. This assists the softening of the fibres.

Another important change produced by the action of heat is the solidification of the albuminous material which forms so important a constituent of lean meat and of other flesh-forming foods. The coagulation of albumen by heat is familiar to every one in the boiling of an egg. The same setting of the albumen occurs at the surface of the meat during the roasting of a joint. In the process of roasting, in which the object is not to extract but to retain all the goodness of the meat, a fierce heat is at first applied all round the joint. This coagulates the albumen near the surface. The joint is then moved rather further from the fire, and cooked at a somewhat lower temperature. A great deal of water evaporates from the meat, but most of the juices are confined in the joint by means of the bag or sack of coagulated albumen. If the joint were kept too near a hot fire after the first coating of coagulated albumen has been formed, the coagulation would go on throughout the joint, the albumen would harden, and the meat would become indigestible. When a well-cooked leg of mutton is cut it is full of juices, which flow out readily. If the joint had been put down before a slow fire, and at some distance from it, the water drawn out of the joint by the heat would have carried with it much of the juice and the fat, which should be kept in by the coagulated albumen.

In those processes of cooking in which the object is to extract the juices of the meat, the coagulation of the albumen has to be avoided.

In cooking starchy food, such as potatoes, the most important change produced by the heat is the swelling up and bursting of the starch granules, inducing a light and floury consistency, a change which is necessary if starch is to be digestible. In order to effect this bursting of the starch granules, a temperature as high as the boiling point of water must be used; hence the rule that vegetables always require, for a part of the time, at any rate, a temperature equal to that of boiling water, but that meat should, as a rule, be cooked at a temperature below that of boiling water.

Section III.—Nitrogen as a Food.

Generally speaking, the flesh-forming foods containing nitrogen are the most expensive. Nitrogen, although so abundant in its uncombined and inert form in the air, generally becomes expensive when wanted for any useful purpose, whether as manure for the land or as food for man.

" The atmospheric fluid, or common air," Lavoisier wrote, " is composed of two gases or aëriform fluids, one of which is capable, by respiration, of supporting animal life, and in it metals are calcinable and combustible bodies may burn ; the other, on the contrary, is endowed with directly opposite qualities—it cannot be breathed by animals, neither will it admit of the combustion of inflammable bodies, nor of the calcination of metals.

"We have given to the base of the former, which is the respirable portion of atmospheric air, the name of oxygen. . . . The chemical properties of the noxious portion of atmospheric air being hitherto but little known, we have been satisfied to derive the name of its base from its known qualities of killing such animals as are forced to breathe it, giving it the name of *azot*, from the Greek privative particle a and ζὼη, *vita;* hence the name of the noxious part of atmospheric air is *azotic gas.*"

Lavoisier occupied himself for many years in developing the knowledge of the chemistry of oxygen—the active constituent of air. The readiness with which oxygen gas can be made to act upon and combine with other substances enables its chemical functions to be determined with comparative ease, and the mode in which this element is used in the nourishment of the animal body was soon elucidated with tolerable completeness ; on the other hand, the inertness of *azot* (or nitrogen, as the substance was called in England) rendered the chemical study of this constituent of the atmosphere a less attractive as well as a slower and more laborious pursuit.

Thus, if we compare the ways in which the oxygen and nitrogen of the air are taken hold of to build up the animal body, we find that one process is direct and in its main features simple, whereas the processes by which the nitrogen of the air becomes a constituent of the flesh and muscle are, on the contrary, indirect and complicated.

When air enters the hollow cells of the lungs, the blood, flowing round the cells in the little veins and capillaries, seizes upon and " fixes " the oxygen (which freely passes through the thin walls of the cells) and carries to every part of the body the combined oxygen which it has seized upon. Parting in its passage through the body with the oxygen which it has seized from the air, the blood acts as carrier of oxygen from the air to the muscle, flesh, nerve, and other parts of the body.

But the blood has no power to seize upon and fix the nitrogen gas which enters the lungs at every breath. The blood has to obtain its nitrogen from the animal and plant food which is taken into the stomach. Animals used by man for food obtain their nitrogen from plants. Plants have to depend ultimately upon the air for their supply of nitrogen.

The plant, however, does not appear to be able to assimilate the free atmospheric nitrogen through its leaves, any more than an animal can assimilate the gas through its lungs. Leaves catch the carbonic acid of the air as the lungs of animals catch the oxygen of the air ; but the nitrogen, it seems, has to reach the sap through the roots—it has to be fixed in some way—before the plant can feed upon it.

Recent researches seem to have proved that on leguminous or pod-bearing plants there lives a class of bacteria
which have the power of feeding directly upon the free
nitrogen of the air. They " fix " the nitrogen, which after-
ward becomes available for plant food. Perhaps this only
happens after the death of the bacteria, when their substance has undergone decomposition, and the nitrogen
compounds have been converted into a soluble form, so that
they can enter the sap in the same way as the rest of the
food derived by plants from the soil.

The soluble form in which it appears that nitrogen is
taken up by the roots of most plants is known as the nitrate
"form" or state of combination, in which nitrogen is combined with oxygen as it is in nitre and nitric acid.

Besides bacteria, associated with leguminous and possibly
other orders of plants, there is another agency which brings
the nitrogen gas of the atmosphere into combination with
oxygen; this agent is electricity. Electric discharges in
the atmosphere cause the oxidation of relatively small
quantities of nitrogen, and the soluble oxides of nitrogen
thus formed are carried down in solution by the rain, thus
adding to the quantities of nitrates in the soil.

Most of the nitrogen at any moment present in the soil
is, however, derived from the substance of preceding
generations of plants. The nitrogen required to be supplied to the soil for the crop of any one year is only the
difference between what is abstracted from the soil by the
plant crop and what is restored to the soil by the death and
decay of animals and vegetables.

The difference between the two amounts may be relatively small for the whole of the earth's surface, and might
perhaps be *nil* if large quantities of nitrogen as nitrate were
not being constantly carried into the sea by rivers.

The nitrogen in vegetable and animal substances is combined with carbon, and nitrogen in this state of combination
may be termed *organic nitrogen.* Nitrogen when in this form
is not directly available as a food for plants.

By the process termed decay in the case of plants, and
putrefaction or decomposition in the case of animals, the
nitrogen is set free from its combination with carbon, and
ammonia or a compound of ammonia is produced. In
ammonia, nitrogen is in combination with hydrogen; and
in this state of combination, is called *ammoniacal nitrogen.*

Free ammonia gas, evolved during the decomposition of
materials containing organic nitrogen, is ultimately brought
back into the soil in a state of solution by means of rain,
owing to the fact that ammonia gas is extremely soluble in
water. Much of the ammonia produced by the decomposition of organic matter meets at once either with water or
with some material with which it can combine, and thus
from the first is retained in the soil.

Plants are able to feed directly upon ammoniacal nitrogen; but in the greater part of the nitrogenous food of
plants the nitrogen is in combination with oxygen, *i.e.* in
what we have termed the "nitrate" condition.

The decomposition of the organic matter of the soil
(which is commonly called *humus*), is effected through the
agency of the *nitrifying bacteria.* They first split up the
organic matter into water, carbonic acid, and ammonia, and
then further assist the oxidation of ammonia to nitric acid.

The conditions favourable to nitrification are that the
oxygen of the air should have free access, that the soil
should be sufficiently moist, but not water-logged, and that
the temperature should be fairly high.

In the presence of an alkali, or alkaline earth in the soil,
the nitric acid forms a salt (termed a nitrate). Where lime
or carbonate of lime is present, soluble calcium nitrate is
produced, and in this form plants obtain much of their
nitrogen. Where potash is present, nitre or saltpetre is
formed, as *e.g.* in dry districts of India, where nitre is found
as an efflorescence on the surface of the soil. Chili nitre,
or Chili saltpetre, is the nitrate of sodium, and occurs in
large deposits in the rainless districts west of the Andes, in
Chili and Peru. In a wet climate such deposits would be

speedily washed away, and carried into the rivers and the sea.

We have referred to the loss of the nitrogen which the soil of continents undergoes on account of soluble nitrates finding their way into the sea. The great quantities of nitrates from Chili and Peru, which have of late years been applied to the fertilization of the land, are a contribution won back from the ocean; the nitrates of the nitrate beds having been formed by oxidation of guano, the *dejecta* of fish-feeding sea-birds. Recently "artificial guano" has been manufactured to a considerable extent from the carcases of coarse fish, caught for the purpose from ships specially employed in connection with this manufacture.

Where crops are cut and carried from the place where they were grown, it is necessary to provide for the restoration to the soil of certain materials, particularly nitrogen, which are thus removed, instead of being allowed to return to the soil through the death and decay of the plants, as would be the case in a state of nature. This is partially effected by returning to the soil the manure from the animals of the farm which fed upon the crops. When the beasts themselves, however, are sent into the towns, they carry away large quantities of the particularly valuable constituents of a fertile soil, such as nitrogen and phosphorus. In a more primitive condition of agriculture, the beasts would be eaten on the farm itself, and the greater part of the nitrogen, etc., would find its way back to the soil whence it came. Facilities for communication and transport, however, and the concentration of the population in towns all tend to make farming, more particularly manuring or soil-feeding, a more complicated matter. This is especially the case where nitrogenous stuff, *e.g.* hay, is sold off the farm, instead of being consumed upon it, and oil cake, or other artificial feeding stuff, is purchased in its place.

Account has to be taken of the albuminoid ratio in deciding upon combinations of artificial foods for the use of stock. Young growing animals retain a larger proportion

L

of the nitrogen supplied to them than the full-grown beasts which are being fatted. In the former case much of the nitrogen goes to build up the muscular flesh, and the manure given by young growing animals is proportionately poorer in nitrogen. On the other hand, full-grown animals, which are being fatted for market, store up chiefly fatty tissue, which contains no nitrogen, and consequently these animals return a larger proportion of nitrogen to the ground.

In feeding different species of animals, account has to be taken not only of the albuminoid ratio, but of what is called the *digestion co-efficient*, or proportion which the food-stuff assimilated bears to the amount taken.'

This proportion is different in the case of different species of animals. In the case of ruminants a large proportion of indigestible fibre in the food is actually necessary. Sheep cannot assimilate more than half of the twelve per cent. of the nitrogenous matter contained in clover hay. Human beings would be able to assimilate scarcely any of the nutriment in hay, the White King, in "Through the Looking-Glass," who took hay when he was faint, being of course an exception. By submitting the hay or grass to preliminary treatment by an ox or sheep, man is able to assimilate the nitrogen and other nutrients contained in hay in the form of beef or mutton.

To sum up, we may say that nitrogen exists as the—

1. *Atmospheric nitrogen*, in which *the atoms of nitrogen are combined with each other.* The bacteria associated with leguminous plants feed upon atmospheric nitrogen.

2. *Ammoniacal nitrogen*, on which plants can, and to some extent do, feed. *Here nitrogen is combined with hydrogen.* The source of ammoniacal nitrogen is the decay of animals and plants.

3. *Nitrate nitrogen*, in which *nitrogen is combined with oxygen.* This is the principal form in which plants obtain their nitrogen. Nitrates, generally speaking, are formed by the oxidation of the ammoniacal compounds produced in

the decay of animal and vegetable matter. In the presence of alkalies in the soil, such as lime and potash, soluble nitrates are formed which supply both nitrogen and alkali to plants. Nitrate nitrogen is also brought into the soil by the action of electrical discharges in the atmosphere.

4. *Organic nitrogen*, in which the *nitrogen is combined with carbon*. This is the form in which nitrogen is taken by animals, either directly from vegetable food—more particularly from green vegetables—or in the case of flesh-feeding animals, sometimes as above and sometimes in the form of lean meat. The organic compounds of nitrogen, after the death of the animal or plant, furnish the food of the *nitrifying bacteria*, which assist in the work of splitting up these compounds, sending off the carbon as carbonic acid, and leaving the nitrogen in the form of a nitrate.

PHYSICS.

CHAPTER X.

Sect. I.—The Liquefaction of Gases.

The common liquids, such as water, rock oil, mercury and so on, can be readily converted into gases; but many of the common gases, for instance, oxygen, nitrogen, and hydrogen, can only be brought into the liquid condition by the use of special methods and powerful agencies.

Low temperature and high pressure are the conditions favourable to liquefaction, and the history of experiments on liquefaction of gases is mainly a record of devices for producing high pressure and low temperature.

Sulphurous acid gas is condensed at the temperature of an ordinary freezing mixture, or at the pressure which can be obtained by a hand-worked piston in a tube or barrel. It had been prepared in the liquid state before the year 1800 A.D. Chlorine was condensed by Northmore in 1805, but his experiments attracted little notice until Faraday had made a speciality of the liquefaction of gases, and the attention of the scientific world was drawn to the subject. Then, as usually happens, forgotten records were found of earlier work on the same lines. The later, but independent, observation by Faraday (1823) of the liquefaction of chlorine is, however, the real commencement of the systematic study of the subject. Faraday has the kind of priority which is of most importance in scientific discovery, that, namely, of being the first to make the subject fruitful, and the first to make its importance generally understood.

Faraday had been experimenting on the solid hydrate of chlorine which separates out in yellowish crystals when ice-cold water is saturated with chlorine gas. Sir Humphry Davy, to whom at that time Faraday acted as assistant, suggested that the crystals should be sealed up in a glass tube, and heated. Davy gave at the time no reason for his suggestion, and Faraday himself at that time did not know what to anticipate from the experiment. The crystals of the solid hydrate were placed at one end of a Λ-shaped glass tube, which was then closed by sealing up the glass in the blow-pipe flame. The crystals being warmed to 60° F., underwent no change, but at 100° F. "the substance fused, the tube became filled with a bright yellow atmosphere, and on examination was found to contain two fluid substances; the one (chlorine water), about three-fourths of the who.e, was of a faint yellowish colour, having very much the appearance of water; the remaining fourth was a heavy bright yellow fluid, lying at the bottom of the former without any apparent tendency to mix with it. At 70° F. the pale portion congealed (*i.e.* the hydrate separated out), although even at 32° F. the yellow portion did not solidify."

Heated up to 100° F., the yellow fluid appeared to boil, and again produced the bright-coloured atmosphere. It was found that by heating to 100° F. the yellow liquid (fluid chlorine) could be distilled from the pale coloured liquid (chlorine water) so as to get them in different limbs of the bent tube. "If, when the fluids were separated, the tube was cut in the middle the parts flew asunder as if with an explosion, the whole of the yellow portion disappeared, and there was a powerful atmosphere of chlorine produced; the pale portion on the contrary remained, and when examined proved to be a weak solution of chlorine in water with a little muriatic acid."

The paper from which the above extracts are taken was read before the Royal Society by Sir Humphry Davy in 1823. In a note at the end of Faraday's paper Davy says that "in desiring Mr. Faraday to expose the hydrate of chlorine to heat in a closed glass tube, it occurred to me that one of three things would happen: that it would become fluid as a hydrate; or that a decomposition of water would occur (forming hydrochloric acid); or that the chlorine would separate in a condensed state." Further on he remarks, "I cannot conclude this note without observing that the generation of elastic substances in close vessels, either with or without heat, offers much more powerful means of approximating molecules than those dependent upon the application of cold, whether natural or artificial, for as gases diminish only about $\frac{1}{480}$ in volume for every degree of Fahrenheit's scale beginning at ordinary temperatures, a very slight condensation only can be produced by the most powerful freezing mixtures, not half as much as would result from the application of a strong flame to one part of a glass tube, the other part being of ordinary temperature, and when attempts are made to condense gases into fluids by sudden mechanical compression, the heat instantly generated, presents a formidable obstacle to the success of the experiment, whereas in the compression resulting from their slow generation in close vessels, if the process be conducted with common precautions, there is no source of difficulty or danger, and it may easily be assisted by artificial cold, in cases where gases approach near to that point of compression and temperature at which they become vapours."

This "bent tube" method was successfully employed by Faraday in the liquefaction of a number of other gases.

In 1822, the year preceding Faraday's experiments on chlorine, Caignier de la Tour had examined the effects produced by heating volatile liquids, such as alcohol or ether in closed tubes. In these experiments the liquid put into the tube was sufficient to half fill it, and the whole of the tube was heated; so that the conditions were different from those of Faraday's experiments. De la Tour observed that up to a certain temperature the liquid continued slowly to evaporate, the bulk of liquid diminishing, and the quantity of vapour increasing. This is the ordinary process of evaporation, which the eye can trace by observing that the position of the *meniscus* which separates the liquid and the vapour becomes lower as the temperature increases.

When, however, a certain temperature is reached the *meniscus* suddenly disappears, and there no longer appears to be any liquid in the tube. On cooling the tube, the inverse change takes place with equal suddenness, a *meniscus* (the line of separation between liquid and vapour) suddenly appears, showing that on lowering the temperature a large amount of liquid is suddenly formed.

De la Tour's experiments point to conclusions quite opposite to those of Davy, quoted above ; since the effect of the "approximation of particles" brought about by pressure was more than counterbalanced by the action of heat.

In 1845, in a second paper on "Liquefaction of Gases," Faraday writes with a mastery of the subject which neither he nor Davy possessed in 1826. He says (Phil. Trans. 1845, p. 155) : "My hopes of success beyond that heretofore obtained, depending more upon depression of temperature than on the pressure which I could employ in these tubes, I endeavoured to obtain a still greater degree of cold. There are, in fact, some results no pressure may be able to effect. Thus, solidification has not yet been conferred on a fluid (*i.e.* a gas), by any degree of pressure. Again, that beautiful condition which Caignier de la Tour made known, and which comes on with liquids at a certain heat, may have its point of temperature for some of the bodies to be experimented with, as oxygen, hydrogen, nitrogen, etc., below that belonging to the bath of carbonic acid and ether, and in that case no pressure which any apparatus could bear would be able to bring them into the liquid or the solid state."

The "bath of carbonic acid and ether" here referred to, was a device for obtaining a very low temperature, a kind of improved freezing mixture. Thilorier, and afterwards

Natterer, had constructed apparatus in which carbonic acid gas could be liquefied by pressure alone. When liquid carbonic acid is exposed to the air at the ordinary pressure, some of the liquid evaporates very rapidly, thereby chilling the lower layers of liquid to such an extent that they freeze, forming solid carbonic acid. If ether be mixed with this solid carbonic acid, and if the pressure be diminished by means of the air pump, the further cooling due to evaporation of the ether, reduces the temperature to $-110°$ C. This extremely low temperature was employed in experiments by which Faraday endeavoured unsuccessfully to liquefy hydrogen, oxygen, and nitrogen.

Andrews employed the same means of producing cold, but used more powerful apparatus for compression. Although by combined cold and pressure the above-named gases were reduced to less than $\frac{1}{500}$ of their original volume, no liquefaction took place. Andrews also conducted experiments upon the phenomenon observed by De la Tour, and showed that above the " critical temperature " of $31°$ C. the greatest pressure which he could bring to bear was not sufficient to liquefy carbonic acid, but that if the temperature were lowered the liquefaction took place at once.

In the later and successful experiments on the liquefaction of the "permanent" gases, such as nitrogen, oxygen, and hydrogen, the skill of the experimenter has been chiefly shown in the means devised for obtaining temperatures below the " critical point " of these gases.

Pictet, of Geneva, who liquefied oxygen in 1877, relied for the production of a low temperature upon the well-known principle of latent heat evaporation. The novelty in his application of this principle lay in the fact that he employed *two* evaporating substances, namely, sulphur dioxide and carbonic acid. By the use of the doubly acting pumps employed in refrigerating machinery, liquid sulphur dioxide can be obtained having a temperature of $-65°$ C. In Pictet's apparatus the cold sulphur dioxide is contained in an annular vessel, which forms a jacket round the tube in which carbonic acid is condensed by the action of another doubly acting pump. The liquid carbonic acid is reduced to a temperature of $-65°$ C. by the cooling action of the "jacket," and, consequently, when the pump is reversed and used as an exhaust pump, the evaporation of the cool liquid produces an extremely low temperature, and the vapour can be again condensed in a jacket round the oxygen tube, at a temperature of $-130°$ C. This tube, of copper, was made very strong to withstand pressure. The oxygen gas was passed into the tube direct from a strong iron retort in which it was evolved, and thus the pressure continued to rise as more and more oxygen entered the tube. At a pressure of about 500 atmospheres, the pressure gauge remained stationary, showing that liquefaction had begun. The whole of the tube was at length filled with liquid oxygen, which was examined by opening a stop-cock, when a jet of a lustrous liquid issued with great force from the tube, to be speedily dissipated by evaporation.

Cailletet, who worked independently at the same problem, first succeeded in liquefying oxygen on the very same day as Pictet. He used simpler appliances, and worked on a different principle. He relied for obtaining his frigorific effect upon the *chaleur de détente*, or latent heat of expansion, of the gas with which he was working. The term "latent heat of expansion," is not a very good one, as a gas is not necessarily cooled by merely expanding to fill a vacuum. When, however, a gas is allowed to expand in such a way as to do mechanical work, the gas loses in heat the thermal equivalent of the mechanical work performed.

In Cailletet's experiments, the gas was contained over mercury in the capillary bore of an immensely strong glass tube. The tube was screwed into an hydraulic press worked by the leverage of a large wheel. The experimental glass tube was surrounded by a freezing mixture of which the temperature for the experiments was not lower than about $-30°$ C.; so that in Cailletet's experiments no attempt was made to surround the gas with a very cold atmosphere. When the pressure attained 300 atmospheres, the oxygen still remained in the gaseous condition, being much above the critical temperature, but on suddenly withdrawing the constraining force from the piston of the press, the gas as suddenly expands, the elasticity or spring of the gas drives back the liquid and the piston, and the sudden mechanical effort of the gas is accompanied by a sudden chill, sufficient to bring the temperature below the critical point. Liquefied oxygen was seen in the tube immediately the pressure was released.

Cailletet's method had the advantage, that the process could be watched through the glass walls of the tube. On the other hand, Pictet's arrangement enabled him to prepare a larger quantity of the material, and to observe its behaviour when exposed to the atmosphere. As we have said, the liquefied oxygen was dissipated immediately by evaporation, so that no further examination of its properties could be made.

Experiments conducted since those of Cailletet and Pictet in the year 1877, have been designed so as to permit of the examination of the physical properties of the liquefied substances.

This has been effected by obtaining such a low temperature of the material, and such a low temperature of its immediate surroundings, that the evaporation of the liquid (whether hydrogen, oxygen, or nitrogen) only takes place slowly under ordinary atmospheric pressure. In Cailletet's experiments a liquid was obtained under low pressure, but the surroundings of the liquid were relatively warm, so that the substance could not long remain liquid. In Pictet's experiments the liquid could only be examined by removing it from the cold surroundings.

In the experiments conducted by Dr. Olszewski, the gas to be experimented on was contained in the innermost of four glass tubes placed one within the other. In the outermost tube was placed solid carbonic acid and ether. By placing this in connection with an air-pump, the temperature of the neighbouring tubes is reduced to −100° C. This was the method for obtaining low temperatures employed by Faraday in his later researches. Now comes a novelty. Ethylene gas, brought from a Natterer's cylinder, is led into the (second) inner tube. Here it is liquefied by the low temperature, under a considerable pressure. The two innermost tubes (into which the oxygen or hydrogen are presently to be brought) are now surrounded by a tube containing the liquid ethylene at about −100° C. This liquid is protected from the warmth of the air by the outermost jacketing tube of carbonic acid ice. By the action of an air-pump the pressure on the liquid ethylene is reduced to 10 mm. of mercury (about $\frac{1}{76}$th the ordinary atmospheric pressure), dry air at the same time is cautiously blown through the ethylene to prevent its evaporation from becoming violent. The gas (say oxygen) is now passed into the two innermost tubes. The intense cold produced by the evaporation of the liquid ethylene, the liquid being to begin with at about −100° C., liquefies the oxygen, which is under a considerable pressure.

One more device remains to be mentioned, the most singular of all in Dr. Olszewski's process. The *two* innermost tubes, as has been said, contain the liquefied oxygen. They are now both put into connection with the air-pump, and the pressure is cautiously diminished. The liquid in both tubes begins to evaporate and is thereby chilled. Presently, the liquid in the outer tube begins to evaporate more quickly than that in the innermost one, owing to the fact that it is in contact with the (relatively) warm ethylene tube. The whole of the liquid in the outer tube consequently evaporates, whilst there still remains a considerable portion of the liquid in the innermost tube. The temperature of the innermost tube has now sunk considerably lower than that of the liquid ethylene. Nitrogen can be frozen in this way. By diminishing the pressure on the solid nitrogen, and thereby causing evaporation, Olszewski obtained a temperature of −225° C., or less than 50° C.

from the supposed absolute zero of temperature, that is to say, the point at which *all* the heat has been extracted from a body.

The exceedingly cold liquid contained in the innermost tube is protected from the relatively warm ethylene by the non-conducting layer of rarefied gas in the intermediate tube. Consequently, the substance remains liquid at atmospheric pressure, or even at lower pressure, for a space of time (5 to 15 minutes) sufficient to allow of an examination of some of the important physical properties. Thus the specific gravity is determined by measuring the height at which the liquid stands in the tube, hence deducing the volume of the liquid; then collecting the gas after evaporation, and measuring its volume. The weight of a given volume of the gas is, of course, well known. This is necessarily equal to the weight of the liquid before evaporation. Hence we know the weight of the liquid in the tube. Its volume having been ascertained in the way described, the specific gravity is readily calculated under atmospheric pressure; it is found that—

	Melts at	Boils at	Critical Temp.
Oxygen ...	—	$-164°$	$- 118·8°$ C.
			Sp. gr. of liquid
			$1·124$ at $-181·4°$ C.
Nitrogen ...	$-214°$	$-194·4°$	$-146°$ C.
			Sp. gr. of liquid
			$·885$ at $-194°$ C.

Hydrogen at $-213°$ C. liquefied under a pressure of $1·90$ atmospheres. The temperatures recorded in the above observations were all registered with a hydrogen thermometer. Liquid oxygen, and, better, liquid air have more recently been employed for obtaining the lowest temperatures, in place of liquid ethylene. By such means Olszewski cooled helium to a temperature calculated to be $-264°$ C., the pressure being 1 atmosphere, without, however, liquefying it. The boiling point of hydrogen he finds to be about $-240°$ C. and its critical temperature

—220° C.   The negative results with helium have served the useful purpose of verifying the indications of the hydrogen thermometer.

SECT. II.—FREEZING MACHINES.

Of late years the production of artificial cold has become an important industry, through the application of principles which have long been known, but of which the practical development presented considerable difficulties. The requisite impulse was given by the need of finding means for preserving meat in a fresh condition during its passage from foreign countries.   For such purposes as this the freezing machines of Carré, which still figure in some of the text-books on physics, are wholly inadequate.   The problem was first practically solved by Coleman by the construction of the Bell-Coleman air-machine, an apparatus so well thought out and perfected that in its first trial a cargo of meat of a value of £8000 was transported across the Atlantic in a perfectly fresh condition.

From 1879 the industry of refrigeration has rapidly increased in importance as new applications have been perceived, and as further improvement in machinery have been effected.   The subject has engaged the attention of many able engineers, and more than three hundred patents have been taken out in connection with it.   At the present time the production of low temperatures is important, not only in the meat trade, but also in the preservation of other perishable articles of food, as fish, eggs, and butter, and in the brewing industry, and in the production of ice.   New applications are being found every day, among which may be instanced the preparation of preserved fruits, and similar processes where crystallisation from solution has to be effected.   The important problem of the cooling of theatres is also receiving attention.

The principles involved in refrigeration present many

interesting features. For the production of *high* temperatures it is usual to employ the agency of chemical
affinity, chemical combination in the process of combustion being attended with an evolution of heat. On
the other hand, many chemical compounds are formed
with *absorption* of heat; but these can only be produced
by the aid of an external supply of energy, and their
formation cannot be employed for the production of low
temperatures.

In order to understand the refrigerating process it is
necessary to consider, in the first place, the means of
attaining high temperatures and the working of heat engines.

By the burning of fuel in the furnace, steam is produced
in the boiler of a steam-engine, and by the changes of
volume of the working substance (steam) a portion of the
energy of heat is transformed into mechanical power. In
heat-engines, of which steam-engines form one class, a
calorific effect is converted into mechanical power, whilst
in refrigerating machinery, on the other hand, mechanical
power is so employed as to yield a calorific effect; but in
this case the calorific effect is *negative*, and the final result
is the production of a *low* temperature.

The refrigerating machine is not in itself a complete
apparatus, since it requires to be *driven* by a steam-engine.
In order, therefore, to attain logical precision in our view
of the artificial production of cold, it is necessary to consider as one complete system the combination of the steam
engine and the freezing *machine*. In this dual arrangement
we start with the production of a high temperature in a
furnace, and finally attain a very low temperature in the
freezing chamber.

The working parts of the freezing machine are very
similar to those of the steam engine. In both there is a
system of cylinders, pistons, and valves, and a working
substance, which undergoes alternately compression and
expansion.

In the Bell-Coleman machine the working substance is

M

air. The process begins with the compression of air by the stroke of a piston in the *compression cylinder.* The power which drives this piston is obtained directly from the piston of the steam engine. The compression cylinder is surrounded by a jacket in which cold water constantly circulates, the heat generated by the compression of the air being almost entirely taken up by the cold water. Thus we obtain air very little above the ordinary temperature, but under a high pressure.

When the pressure is released the air expands. If the expansion be allowed to take place into a vacuum, then, as Joule proved, no change of temperature takes place. But if the expansion take place under such conditions that mechanical power is developed, the mechanical work is done at the expense of the heat of the expanding air, which consequently is chilled. This is what takes place in the *expansion cylinder.* The air, in expanding, drives a piston, which is connected with the cylinder of the steam engine in such a way that it aids the back stroke of the piston in the steam cylinder. Thus frigorific effect is obtained in the refrigerating machine by an action which lightens the work of the driving engine.*

By means of this expansion the air is readily cooled to −50° Fahr., or, if desired, to a still lower temperature. It was here that a great practical difficulty came in. Atmospheric air contains water-vapour, and at such low temperatures this was deposited in the form of hoar-frost. This frost or snow choked the valves and otherwise hindered the working of the machine. It was not found practicable to remove the moisture entirely before the admission of the air to the machine; and until Mr. Coleman's invention the snow

---

* A refrigerating air-machine is the inverse of a hot-air machine. If T_1 and T_2 are the absolute temperatures of the compression and freezing chambers, the efficiency is proportional to $\dfrac{T_2}{T_1 - T_2}$, so that efficiency is greater with small expansion. On the other hand, expansion must be great if great frigorific effect is to be obtained. In practice the efficiency may attain 80 per cent., or rather more.

difficulty appeared to condemn the use of air as a working substance.

The difficulty was overcome by the device of allowing a partial expansion of the air before it entered the expansion cylinder. This preliminary partial expansion is effected in sloping tubes placed in the refrigerating chamber itself. Under these conditions, the aqueous vapour deposits, not as snow, but in a mist or rain, and the moisture is run off by taps placed at the bottom of the sloping tubes. The air thus freed from moisture enters the expansion cylinder to undergo the second and greater expansion by which the principal part of the frigorific effect is obtained.

The cold-air freezing machines are those most used on board ship for transport of meat from Australia, New Zealand, and America. The meat is placed in large chambers, the walls of which are double, the interspace being filled with wood charcoal as a non-conducting material. A jet of intensely cold air is delivered into the chamber at each stroke of the piston of the expansion cylinder, and the temperature of the chamber is thus kept at, or near, the freezing point during the whole voyage.

There is another class of freezing machines, of which the ammonia machines are the type. In this second class, the working substance is not a permanent gas, such as air, but a substance (such as ammonia) capable of being condensed to a liquid by pressure, even at the ordinary temperature of the atmosphere. In these machines the frigorific effect is due, in the first place, to the heat absorbed by the vaporisation of the liquefied substance ; and, secondly, as in the air machines, to expansion of the vapour. Volume for volume, the working substance exercises a much greater cooling effect in the ammonia machines than in the air machines. Consequently, the machinery is more compact and more economical of fuel.

An important difference between these two types is,

that, whereas the air machine works with an *open cycle*, drawing in a fresh supply of material at each stroke of the piston, the ammonia machine works in *closed cycle*, the same working material going through the same round of changes over and again. It will readily be perceived that this circumstance necessitates very different arrangements in the freezing chamber to those which have been described above, where the working substance itself is delivered from the machine and is the direct cooling agent. The refrigerating chamber connected with an ammonia machine is generally cooled by the circulation of a cold liquid in pipes, on a system similar to that employed in heating by means of hot-water pipes. The liquid is some solution having a very low freezing-point, such as a solution of calcium chloride in water, *brine* being the term generally applied to such solutions.

In the ammonia machines a special cylinder for expansion is not required, the expansion being allowed to take place in long coils of tubing, which are placed in a bath in which the brine is kept circulating. From this bath the cold brine is driven by pumps through a system of tubes.

An important advantage possessed by the ammonia machines is the fact that there is no moisture to be removed, and their construction is in consequence considerably simplified. Except on board ship they have undoubtedly an advantage over the air machines, and are coming daily into more general use. For marine "installations"—to use the trade term—the air machines are preferred, owing principally to the fact that, in the case of accident, the working substance could be easily renewed. Moreover, in the case of ammonia, the escape of the working substance, owing to an accident in rough weather, would be highly unpleasant.

An interesting application of cooling by means of brine has been made in mines. One of the greatest difficulties which can occur in the operation of sinking a shaft, is presented by a stratum of sand saturated with water. In more than one case this difficulty has been overcome by freezing the sand and water into a firm mass, and then continuing the sinking operations as if the material were solid rock. The shaft having been sunk to the upper surface of the wet sand, a number of small bore-holes are made to the bottom of the stratum, and in these are placed tubes closed at the bottom, through which cold brine is circulated from a tank at the surface, which is cooled by an ammonia machine. In the course of a few days the wet sand is frozen to a solid mass, and the boring can be proceeded with.

In spite of the varied applications which have already been found for artificial cold, the refrigerating industry must be looked upon as still in its infancy.

CHAPTER XI.

PHOSPHORESCENT BODIES.

ANY substance capable of shining in the dark was originally termed *a phosphorus.* Several substances having the property were known as early as the middle of the seventeenth century, such, for instance, as barium sulphide, the "Bonnonian Phosphorus," and the sulphides of calcium and strontium.

The power of barium sulphide and similar bodies to *phosphoresce* depends upon their being previously exposed to light. When a ray of light falls on any substance, part is reflected and part absorbed. As long as the body is exposed to the ray of light it is itself a source of luminous disturbance, and consequently a visible object. The peculiarity of phosphorescent substances, such as the sulphides of the alkaline earths, is that they continue to be a source of luminous disturbance for a time, when no longer exposed to the ray of light. This power of a body to store up, and slowly dole out, the luminous vibrations it receives is called, in physical optics, *phosphorescence.*

Of all substances luminous in the dark, common yellow phosphorus is the best known, and the example which one would naturally cite as that of the typical phosphorescent body. Singularly enough, the causes which induce the glow of the chemical element phosphorus appear to be altogether distinct from those we have mentioned as the cause of phosphorescence in barium sulphide and similar bodies.

The terminology of the subject has undergone a peculiar alteration since the seventeenth century. At that time any substance capable of shining in the dark was called "a phosphorus." Now, the name phosphorus is restricted by chemists to one chemical element. The element in question exists in more than one allotropic modification, and one of these forms (red, or amorphous, phosphorus) does not shine in the dark.

Ordinary yellow phosphorus was first prepared by an alchemist of Hamburg named Brandt, and in spite of the care with which the secret of its preparation was guarded, a number of persons soon became possessed of the method of manufacture. The early methods, however, gave but small yield, and were so difficult to carry out that the substance remained for long an extremely expensive chemical curiosity. Its many remarkable properties were a favourite subject for exhibition among the learned and curious, and earned for Brandt's production the name of *phosphorus mirabilis.*

Robert Boyle observed that the *phosphorus mirabilis* differed from other shining bodies, in that its luminosity did not depend upon its being previously exposed to light.

Subsequently, it was observed that if phosphorus be brought into the vacuous space above the mercury in a barometer tube, the body no longer shines in the dark. It seemed probable, therefore, that the glow was induced by the presence of air.

As in most phenomena in which air takes part oxygen is the active agent, it appeared likely that the glow was due to some action between the phosphorus and the oxygen of the air. It was found, however, that in pure oxygen phosphorus at the ordinary temperature and pressure did not glow at all. The glow can, however, be induced either by partially exhausting the oxygen in the vessel, *i.e.* by diminishing the pressure, or by raising the temperature. If after the latter means has been adopted the oxygen be compressed, the glow disappears. Now, at ordinary temperature phosphorus volatilises, or evaporates, at a very appreciable rate. It seemed probable that the glow was due to an action between the vapour of phosphorus and oxygen, the two factors essential for the production of the glow being the presence of oxygen and conditions favourable to evaporation.

The correctness of this conclusion is well shown by the following facts. If phosphorus be placed in hydrogen, or in carbonic acid, no glow is seen, but traces of phosphorus vapours can readily be detected in the gas. When one of these gases charged with the vapour of phosphorus is brought into contact with oxygen gas, a glow is at once observed. This glow is stronger in the case of hydrogen than when carbonic acid is used, which is in accordance with the fact that phosphorus evaporates more readily in an atmosphere of the lighter gas.

It appears that the glow of phosphorus in oxygen is in some way connected with the presence of ozone. When a stick of phosphorus is placed in moist air, ozone is produced, and it has further been observed that if a drop or two of ether, or oil of turpentine, substances which destroy ozone, be placed in the ozonised air of a vessel containing a piece of the phosphorus, the glow of the phosphorus is at once quenched. It seems that the glow is nothing else than a very feeble flame, which may be seen when circumstances are favourable to the oxidation, or burning, of the phosphorus. If the temperature be raised to a moderate degree the combustion takes place with greatly increased energy, and we get the ordinary flame of burning phosphorus. Professor Thorpe, and other workers at the Royal College of Science, have investigated a similar case of phosphorescent appearance due to oxidation, which occurs with the *trioxide* of phosphorus.

When phosphorus is burnt in a rapid current of air, one of the principal products is the trioxide, which is capable of combining with a further dose of oxygen, forming the better known and more stable substance, pentoxide of phosphorus.

The phenomena accompanying this oxidation of the lower oxide (the trioxide) have been carefully studied. A glow is observed when oxidation occurs, and it has been found possible, by varying the conditions of temperature and of pressure, to pass insensibly from the feeblest glow to the most brilliant combustion. The trioxide is a more volatile body than phosphorus itself, and better adapted for experiments to show the gradual passage from the degraded combustion of the glow to the ordinary burning with bright flame.

There are other well-known appearances besides those presented by phosphorus, which are due to degraded combustion; one example is furnished by the feeble lambent flame seen inside the wire gauze of a Davy lamp in "fiery" parts of a coal mine. The conducting power of the wire gauze distributes the heat of the flame over a large area, and prevents the inflammable gas outside the lamp from becoming heated to the point at which explosion occurs.

The *ignis fatuus* is another striking example of the degraded combustion of an inflammable gas. Marsh gases, the slow combustion of which is seen in the feeble flame of the will-o'-the-wisp, consist largely of methane, or fire damp, the explosive gas of mines. Their slow oxidation (or degraded combustion) by the air of the marshes produces at night time a faint glow of uncertain or shifting position, to whose misleading light have been attributed difficulties of the road—as many as have beset the search of scientific men after the true cause of the glow of phosphorus.

We pass on to consider the causes of phosphorescence as the term is used in physics.

The following experiment explains how it is that in the case of transparent substances, such, *e.g.* as a solution of copper sulphate, the colour, as seen by reflected and by transmitted light, is generally the same.

The clear solution is placed in a vessel, the sides and bottom of which are carefully blackened, and the mouth of

the vessel is illuminated by ordinary white light.   On looking into the vessel the contents appear, not blue, but *black*.
If a blue tinge be distinguished, it is owing to the sides of
the vessel not being perfectly protected with the required
coating of dull black.   In this case the copper sulphate,
though illuminated by white light, neither gives back a
white light, nor the well-known bright blue reflection, but
reflects practically no light at all.   If, however, a little
finely powdered chalk be introduced into the copper
sulphate, the contents of the vessel immediately reflect the
familiar bright blue colour.

What happens is this: the light falling on the particles
of white chalk is reflected back to the eye of the observer.
In its passage into and out of the copper sulphate solution
the light has been deprived by absorption of the orange
part of the spectrum, and the blue rays alone emerge.
Thus the colour of a solution of copper sulphate is the
same, whether light falls on it or comes through it, because
there is, practically, no true surface reflection at all, but
absorption only.   When light falls on a vessel containing
the liquid, the blue colour is due to light from the background, or from solid particles in the liquid; the orange
part of the reflected light being absorbed by the copper
sulphate, and blue only passing.   We see, therefore, that
the colour of a solution of copper sulphate is always due to
absorption and not to surface reflection.

In the case of a green leaf which shows nearly the same
colour as seen by light reflected from the upper surface as
by transmitted light, the explanation of what goes on
appears to be very similar to the last case.   The light
penetrates some little way below the surface of the leaf, and
is there reflected back through the semi-transparent material.
The chlorophyll absorbs the red rays and allows only the
green to pass out again.   With the leaf, as with the copper
sulphate, there is no true surface reflection.

A very interesting class of optical phenomena is presented by certain substances in which the absorbed rays, or

some of them, are not extinguished, but are modified so as to emerge with a colour different from that which they originally possessed. The colour of a ray depends upon the period of vibration, the violet rays vibrating more rapidly than the blue, and the blue more rapidly than the red, which have the longest period of vibration. The rays of colour of shortest vibration-period are termed the more refrangible rays from the fact that, in passing through a prism or lens, they are bent further out of their original path than the rays of longer period (the less refrangible), such as the red rays. Substances of the class to which we have referred (which are termed *fluorescent* substances) change rays of higher to rays of lower refrangibility, *e.g.* violet to blue. A good example of the action of such substances is furnished by a solution of sulphate of quinine. To a casual observer the solution appears clear and practically colourless. If, however, the eye be placed nearly on a level with the surface of the liquid on which a ray of white light falls, and a black screen be placed behind the vessel which contains the liquid, a bright blue colour is seen on the surface, and for a short distance below the surface, of the liquid. The singular thing about the phenomenon is that this blue colour is not due to the absorption of blue rays, but of rays of another colour, viz. the more refrangible violet rays. This can be shown by examining with the spectroscope the light transmitted by the solution. It is found that it is not the blue but the violet rays which are wanting from the spectrum.

If the white light of the sun be caused to pass through yellow glass the emergent light produces no fluorescence in a solution of sulphate of quinine, the active violet rays having been removed by the yellow glass. Sunlight may, however, be passed through a violet glass without diminishing its power of causing a fluorescence in the quinine.

It is possible to quench entirely a beam of white light by interposing first a violet-tinted glass which absorbs all rays but the violet, and then a yellow glass which quenches these violet rays. If, however, a vessel containing a solution of sulphate of quinine be introduced between the two glasses, light at once shines through. This is because the violet rays falling on the sulphate of quinine are changed to blue rays, which the yellow glass does not absorb.

The properties of fluorescent substances have been applied to the important problem of mapping the ultra-violet portion of the solar and other spectra. The radiation of hot bodies is not confined to visible rays of the ordinary spectrum as perceived by the eye when light is passed through a prism. A thermometer, or a thermopile, placed beyond the visible red is still affected by the radiation which is not visible to the eye; the heating effect of the "ultra-red" rays being very great.

Again, it has been found that there is radiation beyond the visible limit of the deepest violet.

The ultra-violet radiation has but little heating effect, but is potent in producing chemical change.

The fluorescent substance placed in the invisible ultra-violet shines with a blue light readily perceived by the eye. The band of blue light is crossed by dark lines of the ultra-violet part of the solar spectrum. The dark lines are due to absorption of particular rays in the solar atmosphere. Where these lines occur there is no radiation, no ultra-violet ray, and consequently nothing to cause fluorescence. The blue band is therefore crossed by dark lines, the position of which can be measured.

In most cases fluorescence lasts only so long as the substance is exposed to the radiation, but in some substances the emission of light continues after the exciting radiation has ceased to act. *This phenomenon of persistent fluorescence is termed, in physics, phosphorescence.* It is well seen in the behaviour of the sulphides of the alkaline earths, which, after having been exposed to light, continue to shine for a long time in a darkened room.

It remains to inquire into the mechanism of the phenomena of absorption, fluorescence, and phosphorescence. According to the undulatory theory, light is the effect of a wave motion (*vide* Chapter XIII., upon "Æther"). According to the molecular theory of the constitution of matter, substances are made up of small particles, termed molecules, which are in a constant state of vibratory and other motion. The mass of the molecules and the period of vibration is different in the case of different substances. We have to explain what takes place when the wave motion which constitutes light falls on a body composed of such vibrating molecules.

Sir G. Stokes has furnished the explanation in the form of a simple analogy. Suppose a fleet of ships of different sizes to be lying at rest on a calm sea. Suppose a series of waves to pass over the surface of the sea without wind. The waves may be supposed to be the effect of a distant storm. Each ship will begin to oscillate. The time of oscillation, swing, or vibration, will depend upon the size and mass of each ship, and may, or may not, in any particular case be the same as the periodic time of the waves themselves. The duration of a single oscillation of a ship may be the same as that of the wave, or it may be greater. In no case can it be less.

The oscillating ships themselves become centres of disturbance from which waves are propagated over the sea. The periods of these waves may be the same as or greater than those of the original waves, but cannot be less.

The waves from the distant storm correspond to the light waves from a luminous body. The ships of various tonnage correspond to the molecules of different substances. Those ships which vibrate in a slower period than that of the original waves, and themselves cause fresh waves of the slower period, correspond to the molecules of a fluorescent substance. They act like the molecules of sulphate of quinine, which, when agitated by the rapid vibrations of violet light, take up a slower period of vibration, and set in motion in the surrounding ether waves of this slower period, which affect the eye with the sensation of blue light.

Properly speaking, these ships correspond more closely to the molecules of a phosphorescent body, since they would continue to vibrate for some time after the subsidence of the disturbing waves.

CHAPTER XII.

IF we may consider any body which has the property of attracting iron as being a magnet, then the earth itself must be looked upon as the primitive magnet.

Iron is not necessarily a magnet, but every piece of iron has a tendency to become one under the inductive action of the earth's magnetic force. If a piece of soft iron be placed in a vertical position its lower end becomes, in these latitudes, a north-seeking magnetic pole. By hammering the bar while in this position it becomes a permanent magnet.

An iron ship in course of building is subjected to constant hammering, and when it leaves the stocks is a powerful permanent magnet.

Native iron is very rarely met with, but the oxide of iron called loadstone, containing seventy per cent. of the metal, occurs abundantly, and may be considered as a primitive magnet, in the sense that it was the first *known* magnet, and was not produced by man's contrivance.

With the exception of steel and iron, some of the iron ores, and the allied metals cobalt and nickel, all other bodies are *practically* non-magnetic; their magnetic properties being almost infinitesimal in comparison with those of iron.

This circumstance gives a peculiar character to the science of magnetism, and makes the study of the subject very different from that of electricity, which is associated with all sorts of materials.

Another striking difference between magnetism and statical electricity consists in the fact that we cannot isolate "north" magnetism from "south" magnetism, as we do positive from negative electricity. When an iron or steel bar is magnetised, positive and negative magnetism are produced in equal strength and the magnet has its north and south pole. One cannot draw off the "south" magnetism and leave a "north" magnet. The experiment may readily be tried on a watch spring which has been magnetised, after having been heated to redness and cooled suddenly. The watch spring can now be easily broken, and each piece is found to have its north and south end after every occasion of breaking the spring. The north and south ends may be tested by bringing them near to the end of a light and delicately suspended magnet, which readily moves if attracted or repelled.

Since the north and south poles of the magnet act in opposite ways and are always of equal strength, it is evident that the further apart the poles of a magnet are the more likelihood there is of obtaining definite and simple effects, such as can readily be understood.

The use of the common horse-shoe magnet is apt to confuse the learner, although convenient when the object is to attract powerfully a piece of unmagnetized iron, which is attracted both by the north and by the south pole. Hence the convenience of the horse-shoe form, *e.g.* for setting the index of a registering thermometer. A "keeper" of soft iron which prevents leakage of magnetism is also more conveniently employed with the horse-shoe form ; but for those who wish to study the elementary phenomena of magnetism the bar magnet is much better.

The diminution of force between two magnetic poles is proportional to the *square* of the distance between them, the same law of decrease as in the case of the force of gravitation between two bodies. The significance of the law lies in the simple geometrical deduction that no force is lost in transmission. In the case of light, which follows the same law of decrease, the truth of this deduction is readily illustrated by a geometrical figure showing that those rays from a luminous point which illuminate a surface of one square inch at a distance of one foot, will illuminate a surface of four square inches, at a distance of two feet. As the *intensity* of illumination in the second case is found to be one quarter of the intensity in the first case, it follows that the quantity of light which falls upon the second surface is *the same* as the quantity which falls upon the first surface.

This law of " inverse squares " was shown by Coulomb to hold in the case of two magnetic poles.

Coulomb employed his well-known torsion balance for the investigation. In this instrument the magnetic force is balanced against the force of elasticity in a twisted thread or wire. A long and light magnetic needle is suspended at the end of a fine wire, or thread, which is fastened at its upper end to a cap, or head, which can be twisted through any angle, and is provided with a graduated circle to measure the angle through which it is turned.

The angle through which the magnetic needle (and with it the lower end of the wire) is twisted can also be measured by a second graduated scale.

The instrument having been arranged so that there is no twist in the wire when the magnetic needle is in position, the north pole of a bar magnet is brought near the north pole of the needle. The north pole of the needle retreats, twisting the wire until the elastic force in the wire, *i.e.* its tendency to untwist, balances the repellent force which is acting between the magnetic poles.

The graduated scale shows that the repulsion of the magnet has twisted the lower end of the wire through a small angle of, say 6° to the right. By twisting the cap or head, from which the wire is suspended, *to the left*, we can increase the untwisting force in the wire, and thus drive back the north pole of the needle nearer to the north pole of the magnet.

N

Let the left-handed twisting of the cap be continued until the distance between the needle-pole and the magnet-pole has been halved. The lower end of the wire is then, of course, twisted through 3°, instead of 6°, to the right.

It is found that under these circumstances the cap, and therefore the upper end of the wire attached to it, have been twisted through 21° to the left. The total twist in the wire is therefore 24° when the distance between the repellent poles has been halved, instead of 6° as at first. Generally, the twisting of the wire has to be increased *fourfold* in order to balance the force of repulsion at *half* distance.

Now Coulomb had found by experiment that the angle through which a wire is twisted is exactly proportional to the force required to keep it twisted.

Therefore in the above example the *force* between the poles was four times as great at half distance, or the force diminished as the *square* of the distance.*

In a magnetised sewing needle the mass of metal is small, the magnetic charge is necessarily small also; and the poles are well out of each other's way. It is, therefore, a very convenient instrument for observing the simpler phenomena of magnetism.

The sewing needle, having been magnetised by drawing it across the pole of a bar magnet, is laid gently on the surface of still water.

The first thing noticeable is that the needle generally twists about its middle point, after the hand has left it free, and then rests in its new position. If a second magnetised needle be placed in another basin, it will be seen that *both* needles point in the same direction. They both lie in a northerly direction, known as the magnetic meridian.

* *Vide* Coulomb, "Collection de Mémoires, relatif à la Physique," vol. i. The experiment is difficult to carry out. Coulomb's results were approximate only, the errors being from 4 per cent. to 10 per cent. It must be recollected that the force exercised at a given distance depends upon the nature of the intervening substance. The above experiments were done in air.

The effects above described, due to the earth's magnetic force, are different from those produced by bringing the pole of a bar magnet near the floating needle. In that case the needle not only turns to the magnet but follows it, whereas the earth's magnetism give the needle a twist but no pull. If there were such a pull the needle would readily show it, for the resistance of the water is less to any motion lengthwise than to the turning of the needle about its middle point.

It is sometimes said that the earth acts somewhat as if there were a great bar magnet about half the length of the earth's axis buried inside it, making a small angle with the axis of the earth's rotation. Such an analogy is a very rough one, and must not be pushed too far. However, it represents the facts sufficiently well to be employed as a means of explaining the general character of these elementary phenomena. It is evident that the pole of such a magnet, which is by hypothesis situated at a great distance, could only exercise a twisting, not a pulling, force on a magnetised needle; the two poles of the needle being *practically* at the same distance from the pole of the great magnet, the attraction and repulsion on the north and the south poles of the magnet respectively will be equal in amount. A push at one end of the needle and an equally strong pull at the other cannot drag the needle along, but, if the needle be set across the line of force, the push and the pull will act together so as to twist the needle into the direction of the line of force, the magnetic meridian.

In the same way the magnetised needles attached to the under side of the card of a ship's compass, keep the card always in the same position, so that, as the ship's head turns, the compass box revolves round the stationary card, and the lubber line marks the point of the compass towards which the ship's head is turned. The card to which the compass needles are attached has in its centre a brass socket, with a piece of agate at the bottom, which is supported by an upright pin. A slight amount of friction is all that opposes the tendency of the needle to keep in the magnetic meridian during the turnings of the ship's course.

This arrangement does not show, however, whether the earth's directing magnetic force is *horizontal* or not. The needle and the card are free to turn horizontally, but a downward or upward twist, unless very powerful, would expend itself without visible effect against the rigid parts of the apparatus. In the same way, with the sewing needle floating on water, the repulsion between the *greasy* surface of the needle and the water surface is sufficient to resist the downward pull of gravity, and would prevent us from observing a downward twist unless it were a powerful one.

To show the true line of action of the earth's directing force upon a magnet, we must use a " needle " mounted on a *horizontal* axis, and free, except for the earth's magnetic action, to turn in a vertical plane. The form of needle used for showing the magnetic *dip* is the long lozenge shape. If the dip needle be placed so that a line joining the supports be in the magnetic meridian, the needle hangs vertically, and, if the apparatus be gradually turned round, the lower end of the needle tends to rise, until, when the needle is in the magnetic meridian, the dip, or angle which it makes with the horizon, is in these latitudes about 67°. Thus in England the line of the earth's magnetic force slopes steeply downwards in a northerly direction.

The magnetic poles of the geographer are positions on the earth's surface where the dipping needle points vertically downwards. The direction in which such a needle points would meet the direction in which a Greenwich needle points at some thousand miles down in the bowels of the earth. This shows that the centres of the earth's magnetic action are deep-seated.

The term magnetic poles for places on the earth's surface is somewhat misleading. These positions are not "poles" in the sense in which the term is used in connection with a bar magnet or a magnetic needle. In these cases we mean by the "poles" positions near the ends, where there is greatest concentration of magnetic strength, centres of force in the magnet. The dip of 67° in these latitudes shows, as has been said, that the centres of magnetic force in the earth are deep-seated. If there be "earth poles" comparable to the poles of a bar magnet, they must be even further from the surface of the earth than the point at which the direction of the Greenwich dipping needle meets the direction of the needle where the dip is 90°.

The strength of the earth's magnetic force has been measured in the same way as other forces, in terms of the weight exercised by a given mass under the action of gravity.

The determination of the earth's magnetic force requires two principal sets of observations.

In the first, the earth's action upon a magnetic needle is *opposed* to the strength of a bar magnet. We find in this way how many times stronger is the action of the magnet on the needle than is the earth's action. If we call H the force exerted by the earth (so much of it as comes into play in a horizontal direction) and M the strength of the magnet, then our experiment determines the numerical value of $\frac{M}{H}$.

In the second series of observations, we allow the earth's magnetism to act with and reinforce the strength of the magnet, and we determine the value of the two forces acting together and reinforcing one another. Our numerical result is now in the form of a *product*, viz. the strength of the magnet multiplied by the horizontal component of the earth's magnetic force, or M × H.

The method employed is to suspend the same bar magnet as that used in the last experiment by a fine wire or thread, the magnet being placed in a stirrup, with its axis horizontal. Matters are so arranged that when the fibre is without torsion the magnet lies in the magnetic meridian. The magnet is then twisted round its point of suspension, twisting the wire at the same time. On letting go the magnet, we observe the rate of oscillation as the fibre twists and untwists. The rate of oscillation is quicker than that of an unmagnetised bar under the same conditions, for the magnetic force acting between the earth and the poles of the magnet assists the elasticity of the wire to overcome the inertia of the iron bar. This acceleration of the oscillation gives the value of M × H.

In the former experiment we determined $\dfrac{M}{H}$. The value of H can now be calculated, since, if we divide the value MH (obtained when earth and magnet force act together) by $\dfrac{M}{H}$ (obtained when the earth acts against the magnet) the quotient is the *square* of H.

As we know the angle which the line of action of the earth's force makes with the horizon (from the *dip* observation) we can readily calculate the total magnetic force of which H is the horizontal component.

In some observations in the north of England, of which the data are before us, the dip was $67\frac{1}{2}°$. Hence the total force would be about twice the horizontal component. In the experiments to which we refer, it was found that H was o·15 *dynes;* therefore, the earth's magnetic force was about o·30 *dynes.* The dyne is the unit of force of the centimetre-gramme-second system of units, and is equal to about o·016 of the weight of a grain in the latitude of London; o·3 dynes is, therefore, equal to about one two-hundredth of a grain, and o·15 dynes, the value of H, to about one four-hundredth of a grain. This small force keeps the needle true to the pole.

CHAPTER XIII.

THE æther of modern science stands in much the same relation to the hypothetical "media" of the sixteenth and seventeenth centuries, that the theory of chemical atoms having specific weights bears to the older philosophic notions of a "corpuscular" constitution of matter.

Information as to the existence and properties of æther was first gained by the minuter study of the phenomena of light, which showed also that the older theory of the emission of luminous particles failed when applied to the most critical cases.

If light emanating from a luminous point be caused to travel by two paths of slightly different lengths, and the two rays be allowed to overlap upon a screen; then in certain positions the brightness is greater than that of either ray, *and in other positions the screen remains dark.* Two equal portions of light may give a double brightness or may give *no* brightness.

There cannot be a *substance* of which two equal portions produce no substance, neither can there be two substances which annihilate one another. The phenomenon is evidently that of some *condition of substance,* a condition which admits of being completely reversed, such as *motion.*

If two motions, equal and oppositely directed, be communicated to a body, the body will have no motion.

Light appears to be a mode of motion of a substance.

Consider light emanating from a luminous object and illuminating a screen. If, at any position between the screen and the luminous object, a second screen be placed, that in like manner will be illuminated. The substance of which light is a mode of motion (the light-bearing or luminiferous substance) exists at all positions between the luminous body and the first screen, and at each of these positions the substance is in motion, because at each position there is light.

Obviously, the luminous object is the source of disturbance of the luminiferous substance, since if the luminous object be removed or quenched, there is no light. The luminous object is always a hot body, or one which is illuminated by a hot body. The particles of a hot body are in rapid motion: the luminous body moves the adjacent portion of the luminiferous substance. This motion is transmitted by each portion of the luminiferous substance to the next portion, until it reaches the distant screen. As long as the luminous body remains the screen will be illuminated.

The rate at which the disturbance imparted by the luminous body is transmitted by the luminiferous substance is the "velocity of light." This velocity is so great that refined observation is needed to detect the interval of time required for transmission from point to point.

The methods of measuring the velocity of light are described in the text books : the results obtained show that the luminiferous substance can transmit a disturbance at the rate of 186,000 miles per second, and that this velocity is maintained beyond terrestrial limits, in the interplanetary space.

The fact that two portions of light can produce either greater or less illumination than either portion singly, shows that the movements of any part of the luminiferous substance are of the nature of a vibration ; the parts of the substance are *swinging*, not moving continuously in one sense or direction.

We shall learn something about the nature of the luminiferous substance by comparing the velocity of light with the velocity with which ordinary substances transmit a vibration.

By striking a solid body vibration is set up, and the rate of transmission of the vibration through the solid can be measured. Each solid has its own particular rate of transmitting a vibration, but every solid, *e.g.* metal or glass, transmits vibration many thousand times more slowly than light traverses a transparent solid, such as glass. Therefore when light is transmitted through glass, for instance, glass is not the luminiferous or light-bearing substance. The luminiferous substance is in the glass, but is not the glass itself.

A vibrating solid will set a gas vibrating, *e.g.* vibrations of air may be set up by striking a bell. The velocity of transmission of vibrations in air or any other gas (the velocity of sound in the gas) is many thousand times ess than the velocity with which light traverses a gas. Therefore, when light passes through a gas, the gas is not the luminiferous substance; the luminiferous substance is in the gas, but it is not the gas itself.

If a bell be struck under water, vibrations are transmitted through the water, but the rate of transmission of vibrations by water or any liquid is many thousand times less than the velocity with which light traverses a liquid. Therefore the luminiferous substance is in the liquid, but is not the liquid itself.

In every transparent body, and in the spaces between the stars and planets, there is a luminiferous substance.

Whether the substance is contained in opaque bodies the phenomena of light are hardly capable of showing. As far as light phenomena are concerned, we might either suppose that the substance ("æther") does not exist in opaque bodies, or that the substance is hindered from transmitting vibrations in opaque bodies. As light can penetrate a short distance into "opaque" bodies (as is shown by the transparency of thin slices), the latter supposition is the more probable, even on the evidence of light phenomena only. We shall see later on that the conclusion that "æther" exists even in opaque bodies is confirmed by the evidence of electrical phenomena.

The next important step in our knowledge of the lumini-ferous substance, æther, is gained from the inquiry as to the relation between the plane in which the parts of the substance vibrate and the direction in which the disturbance is transmitted.

The fact that vibrations are transmitted at all indicates that *æther has inertia and elasticity*, that when a part is dis-placed it tends to spring back.

Fluids tend to spring back when *compressed;* in solids, like iron and glass, there is a spring-back, even when one part is made to *slide* without compressing the body.   This is *rigidity*, or elasticity of shape.   Only in bodies which have rigidity, or elasticity of shape, can vibration take place at right angles to the direction. in which the vibrations are transmitted.

It has been proved, as we shall presently describe, that æther can transmit a transverse vibration, and it follows that æther possesses rigidity, which is the most characteristic property of a solid.   As we are concerned now with æther rather than with the details of the study of light, it will not be necessary for us to prove that the vibrations which constitute light are *always* in a plane at right angles to the direction of transmission or propagation.

The *interference* of light, which was described above, shows that light consists in the motions of a substance, and that the motions can be contrary in sense.   We have shown that the motions are of the nature of a vibration. Under what conditions, if any, would it be *impossible* for two parallel beams of light to interfere, the total amount of light (*i.e.* the amount of motion) at any position *never* being less than the amount of light of either beam ?

Evidently this can only be if the motion of the parts of æther in one beam at least be *never* either wholly or partly in the direction of transmission.   In other words, the vibration in one beam at least must be in a plane at right angles to the direction of transmission.

To make the matter easier to follow, we will illustrate the simplest case in which interference is impossible. The simplification introduced for the purpose of illustration is, however, not *required* in the argument given above, which applies generally.

Suppose, then, for illustration, that the motions of the parts of the æther in *both* parallel rays are at right angles to the direction of propagation. Then, if the vibrations be rectilinear and at right angles to one another, there can by no possibility be such interference as to produce a diminution of the light, since neither motion is at all in the same sense as the other. If we imagine the two parallel, or coincident, rays of light to pierce this sheet of paper at right angles to the plane of the paper, two straight lines in the plane of the paper, and at right angles to one another being the directions of vibration, then the two rectilinear motions might conceivably compound together so as to give a circular or elliptical motion, *but the amount of motion* (and therefore the amount of light) cannot be diminished. Hence there could be no interference of light in this case.

We see, therefore, that if by any means we can produce two parallel or coincident rays of light which shall be incapable of interfering with one another so as to diminish the amount of light, then we shall know that the parts of luminiferous æther can vibrate at right angles to the direction in which the vibration is transmitted, and we shall conclude that æther has rigidity, the property most characteristic of solids.

Treatises on physical optics describe the arrangements for producing parallel rays of plane polarised light with the planes of polarisation at right angles. Such rays do not interfere, therefore the luminiferous substance is possessed of rigidity.

The rigidity of æther is a factor of such capital importance in the nature of the substance, that we will at this point

enter more fully than we have yet done into the consideration of the property of substances which can transmit such
vibrations as these described above, and we will also deal
more fully with the mechanism of the transmission itself.
In the course of this inquiry we shall see how it is possible
to compare quantitatively the rigidity and the density of
æther with the rigidity and density of ordinary solids.

We may divide materials into two principal classes,
according to the forces evoked by the displacement of a
part.

In the first class (fluids, whether liquid or gaseous) *compression* evokes a resisting force which is transmitted
throughout the fluid, but if one part be caused to slide
over another, which does not change the volume of the
body, no resistance is evoked throughout the fluid. It is
owing to this fact that fluids possess no proper shape.

In the second class, solids, if a portion of the material be
caused to slide, a resistance is evoked, not only in the plane
of displacement, but also *in a direction at right angles to the
plane of the displacement.*

Consider the case of a thin round iron bar or wire.

Let the bar or wire be lying on a table, and let the end
A be twisted in the direction of the arrow. The twist may

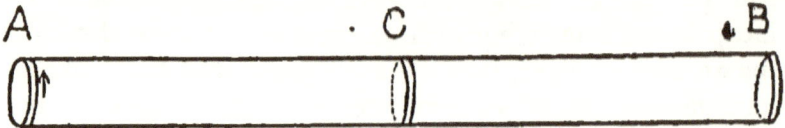

be regarded as an endeavour to cause the portion A to
slide round. Common experience tells us what will
happen. The *whole bar* will turn round. The force
applied tangentially at A will be transmitted longitudinally
so that a tangential force will act at the end B. The
rigidity of the wire is so great that we shall be unable, by
twisting the part A as in the above experiment, to make the
part A *slide* appreciably.

The rigidity of the wire would in like manner prevent us

from pushing aside the part A without moving the rest of the bar. Rigidity enables a solid to resist change of shape.

In order to make A slide appreciably, the most convenient method is to subject the end B to such powerful constraint, that the transmitted force shall not be able to rotate the portion B. For this purpose the end B may be fixed in a vice. It will then be most convenient to let the wire hang vertically from the fixed end B. By applying a sufficient twisting force at A the portion A is caused to slide round.

We have next to inquire what determines the rate at which the transverse sliding motion of A is transmitted along the wire, bearing in mind that *æther* transmits a transverse disturbance at the rate of 186,000 miles per second.

An iron wire or bar is homogeneous, the properties of any piece are the same as those of any other piece.

A homogeneous solid may either be isotropic or not isotropic. In any isotropic solid the properties are the same *in any direction* about any point in the body; for instance, when a portion of the body is displaced, the forces evoked are of equal amount in every direction. Experience and experiment show that iron is isotropic, or very nearly so.

Crystals are homogeneous solids, but many crystals are not isotropic. When a portion of a non-isotropic crystal is displaced, the forces evoked are not of equal amount in every direction. As long, however, as we are dealing with an iron bar, we need not consider the properties of non-isotropic bodies.

We return to the experiment of twisting the end A of the iron wire, the end B being firmly fixed.

When the part A is twisted, a force of restitution, tending to untwist, is evoked. The portion A tries to regain its former position, and its (tangential) force of restitution or recoil, is equal to the force which is transmitted (longitudinally) along the wire, since the body is an isotropic solid in which equal forces are evoked in all directions.

It was experimentally proved, by Coulomb, that if the end A be twisted round *and then let go*, A recovers its original position in a certain definite fixed time. The time of untwisting is the same, whether the end A be twisted through 90°, or through 180°, or through a complete revolution of 360°, or even through several complete revolutions. The force of recoil is evidently very different in the different cases, the greater the displacement the greater the force evoked.

Now forces are compared with one another, by the amount of motion which they produce in equal times. The force of untwisting when the displacement or twist was 180°, is therefore *twice* the force of untwisting for a displacement of 90°, since the original position is regained in equal times in each case.

We have therefore the important law that *the force of restitution is proportional to the displacement.* This law of solids is an ultimate fact, discovered by experiment and, as yet, unexplained.

We are dealing with an isotropic solid; therefore, however much the part A be twisted, the resistance to twisting (*i.e.* the force of restitution), will increase proportionately *in each portion* of the wire. A tangential displacement or disturbance can therefore only be transmitted longitudinally from portion to portion of the wire at a definite speed. If we begin to twist A, C will begin to twist after the lapse of a certain definite interval of time. C will always be a certain fixed time behind A, whether in twisting or untwisting, *any* tangential disturbance being transmitted at the same rate.

Similarly æther is found to transmit tangential disturbance at the velocity of 186,000 miles per second, whether the disturbance be imparted to it by the white-hot filament of an incandescent electric light, or by a red-hot poker, or by a dark-hot Leslie's cube.

Different solids, as can readily be shown, transmit a tangential disturbance at very different rates. Thus, if we take equal lengths of indiarubber cord and of iron wire, fixed at one end and free at the other; and if we twist the free end, we find that the free end of the rubber cord may be considerably twisted before any movement is seen at the half-way point. The iron wire is much more *rigid* than the rubber, and, if a portion be displaced through a distance 1, the force transmitted through the wire is much greater than in the case of indiarubber. Consequently the disturbance is more quickly felt at a distance.

Were the iron no *denser* than the rubber, the disturbance would be still more quickly felt at a distance; but before the disturbance can be felt at a distant point, each intermediate portion must be set in motion, and this requires more time in the case of a denser material.

Let us call the force transmitted in any material, when a part is displaced through a small distance 1, the *rigidity* of

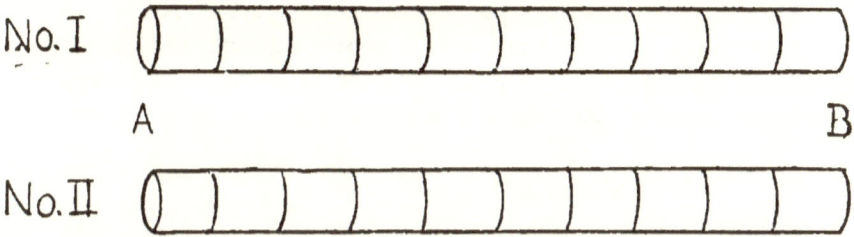

No. I

A   B

No. II

the substance, and let e be the symbol of rigidity. It evident that a disturbance will be transmitted more quickly the greater is the value of e, and more slowly the greater is the density, d.

We proceed to investigate the exact relation between the velocity of transmission, and the rigidity and density of a substance.

Let there be two rods or wires, No. 1 and No. 2 of the same length and size, but of different materials.

Suppose each rod to be composed of an equal number of parts, as shown in the figure. As the size of the parts is the same in each rod, the masses of the parts are proportional to the densities of the two materials.   Before a disturbance can be felt at the far end, B, of either rod, each intermediate portion must have been moved.   The time t taken to move a part through a small distance 1, can be calculated in terms of e and of the mass of the part, which is proportional to the density d.

It must be borne in mind that we are dealing with a body originally at rest, which is set in motion by a force which continues to act upon it.   It is almost the same as the old problem of the motion of a falling body.

Suppose the rigidity of the second rod to be sixteen times as great as that of the first rod, and the density nine times as great.   The acceleration, being proportional to the force, and inversely proportional to the mass, is $\frac{16}{9}$ times as great in the case of No. 2 as in No. 1.   At the first instant the part was stationary.   Each instant the part of No. 2 gains $\frac{16}{9}$ as much speed as the corresponding part of No. 1; in how much shorter time will it move through the same, small distance?   *The times will be inversely proportional to the square roots of the accelerations,* viz. as $\frac{3}{4}$: 1, or, generally, as $\sqrt{\dfrac{d}{e}}$: 1, d being the density and e the rigidity of No. 2, the rigidity and density of No. 1 being taken as unity.*

Looking at the figure, we see that the number of pieces

---

* Compare the familiar problem of a body falling under the action of gravity.   Let g be the acceleration due to gravity.

velocity at starting $= 0$
velocity at end of 1 sec. $= g$
velocity at end of t secs. $= gt$

$\therefore$ average velocity during the first t secs. $= \dfrac{0 + gt}{2}$

$\therefore$ distance (s) traversed in first t secs. $= \left(\dfrac{0 + gt}{2}\right) \times t = \frac{1}{2}gt^2$

Now as $s = \frac{1}{2}gt^2$, $t^2 = \dfrac{2s}{g}$; $\therefore t^2$ varies as $\dfrac{1}{g}$, and t varies as $\sqrt{\dfrac{1}{g}}$.

Thus when a body moves under the influence of a continually acting force, the time of traversing a given distance varies inversely as the square root of the acceleration.

in each rod is the same. The time of transmission along the whole length of the rod is therefore proportional to the time required to set each portion in motion, proportional that is to $\sqrt{\dfrac{d}{e}}$. The velocity is, of course, inversely proportional to the time, therefore—

$$\text{velocity of transmission} \propto \sqrt{\frac{\text{rigidity}}{\text{density}}}$$

The great velocity of light shows that in æther the ratio of rigidity to density is enormously greater than in any other known substance. We can give some notion of the actual value of the density of æther, and hence of the rigidity also, in the following manner.

The amount of energy in a cubic foot of the sun's rays may be measured by determining the heating effect per second, when the sun's rays are completely absorbed by a surface of lamp-black, the energy of the vibrations of æther being converted into those irregular motions of chemical matter which constitute heat.

If the amplitude of vibration, *i.e.* the amount of sideway swing, of the æther in the rays were known, this determination would enable the density of the æther to be calculated from the above determination of amount of energy.

Unfortunately we can only guess at the approximate amount of amplitude which circumstances seem to render probable, and from this a guess is made at the actual density of the æther; a guess which does not pretend to be very accurate, but which is not likely to be altogether wrong. It gives the "order of magnitude" of the density. If the density of water be 1, that of æther is *about* 10^{-22}, the density of air at the ordinary temperature and pressure being about 10^{-3}. Æther is therefore about 10^{19} times less dense than ordinary air.

Assuming this value for the density, the rigidity of æther works out at about 10^{-1}. The rigidity of steel is about 10^{12}, and its density compared to water about 10^1.

O

The disparity between the densities of steel and of æther is much greater than the disparity between their rigidities. Hence vibrations in steel are relatively slow.

The intensity of light moving through a perfectly transparent substance, such as air, diminishes as the square of the distance from the luminous object. The area of the surface which, at any moment, constitutes the *front* of the wave of light, increases as the square of the distance from the source of light. The total amount of light therefore does not diminish in the process of radiation; the vibrations of æther are transmitted, as nearly as observation has determined, absolutely without loss. There is in the æther no evidence of the frittering away of the energy of vibrations into irregular movements, such as would occur through *frictional* resistance. As far as these observations go, *æther appears to be perfectly free from viscosity.*

We return to the consideration of the absorption of the sun's rays by a surface of lamp-black.

Not all the sun's rays which fall upon the surface of lamp-black are visible to our eyes.

Let us call the distance between two disturbed portions of æther in a ray which are simultaneously in the same phase of motion the "length of a wave," or the "wave length."

The eye cannot see by the aid of waves which are too short, just as the ear cannot hear air-waves which are too short. Radiations too short for the eye to perceive will, however, affect many chemical substances; they are the most potent in producing photographic effects. Æther-waves which are *too long* to excite the sensation of sight, can be felt by the eye by their warming effect. The long and short waves appear to be exactly similar to those which excite the sensation of sight, they travel at the same rate, and are capable of polarisation.

The time of the vibration or swing of a disturbed part of æther in radiation of considerable wave-length such as red light, is much longer than the time of swing in radiation of short wave-length such as violet light. A body at a "dark-red heat" sends out radiations of which the time of swing is greater than in the case of radiations from a body at a bright-red heat.* It must not be forgotten that the rate of transmission of disturbance is the same in both cases; the distance between two parts which are simultaneously in the same phase (the wave-length), being proportionately greater in the case of a slow-swinging radiation.

The colder the body is the slower will be the swing it gives to the adjacent portion of æther, and consequently the longer will be the wave-length. Bodies, solids and liquids at all events, at the ordinary temperature of the air, or even at a much lower temperature, send out radiations; a fact explained by the well-known "theory of the exchange of radiations," which is dealt with in treatises upon the science of heat. The time of swing in such cases must be very much slower than in the case of radiation from a red-hot body.

We see, then, that the æther is capable of taking up and transmitting without loss, either a very rapid swing or a relatively very slow swing, from any solid or liquid body, at a high temperature or a low temperature.

We may now inquire whether there is any other mechanism which will set up a vibration or regular swing in æther, besides movements of the small parts of bodies connected with the temperature of those bodies.

This brings us to the study of electrical phenomena.

The discharge of a Leyden jar is an oscillatory process accompanied by a surging to and fro of the electrical charges of the two coats of the jar. The oscillations are more rapid, that is to say, follow one another more

---

* If a difficulty be found in reconciling this statement with the circumstance that in a twisted wire the recovery is always achieved in the same time, the reader must call to mind that by the time of swing is meant the time taken to pass *once* through all phases, whilst in the case of a wire which has been twisted through several complete revolutions, the whole succession of positions or phases is traversed *several times over* before the wire is completely untwisted.

frequently, in the case of a small Leyden jar than in the case of a larger one. Special small electric oscillators have been constructed to give very rapid surgings.

If the æther be capable of being set swinging by these surgings, the "wave-length" will be shorter in proportion as the time of swing is shorter. The velocity of the transmission of the disturbance will, of course, be the same as the velocity of transmission of light, *if æther be the substance, which transmits the disturbance originated by the electric oscillator.*

The vibration-time of the electric oscillator can be ascertained. If a periodic, or oscillatory, disturbance travelling with the velocity of light be set up, the wave-length can be calculated from the frequency or vibration-time.

By using the most rapid electric oscillators the corresponding wave length becomes manageably small, a few feet only.

The next question is, how to seek for an effect which shall show whether the æther is vibrating. It is no use to *look;* for the wave-length would be many times greater than would excite the sensation of light. Perhaps a heating effect might be detected ; but the simplest and best plan is to look for an electrical effect. The case is analogous to that of the ingenious reversible instruments which have come into use of late years, such as the telephone, and the phonograph. We speak into the mouthpiece of the telephone, a disturbance is excited in the wire (no matter for the argument of what kind it be as long as we remember that *it is not a sound*), and this disturbance reproduces the original sound in the ear-piece, which is similar to the mouth-piece. In like manner with the phonograph, sound produces permanent indentations of a prepared surface. The instrument when worked backwards reproduces the sound.

In the same way, if we wish to detect oscillations of æther, we shall invite them by means of a similar instrument *to reproduce the disturbance which produced them.*

A properly devised electrical arrangement will show, by sparking, in what positions the disturbance, if any, is at its maximum. The distance between the sparking positions is proportional to the wave-length.

Such an arrangement was adopted with complete success by Hertz, who showed that a periodic disturbance was set up, and who measured its wave-length. The period of the electric oscillator being also known, the velocity of transmission was readily calculated, and was found to be the same as the velocity of light.

It is therefore concluded that æther transmits periodic electric disturbances, for if it be not æther, but some other substance, then we should have to suppose space to be filled twice over with media of which the ratio of rigidity to density is the same.

The electric waves can be polarised, so that they are produced by transverse disturbances, as is the case with light.

The electric disturbances sent out by the working of many ordinary electric appliances are of very great wave-length—100 miles would be quite a moderate value—whereas the vibrations of light are as short as, say, $\frac{1}{100000}$ of an inch.

The fact that æther can transmit indifferently disturbances of such vastly different periods and wave-lengths, indicates that if æther have a grained structure, as chemical matter has, if there be æthereal atoms, these atoms must be so small that $\frac{1}{100000}$ of an inch and 100 miles are of the same order of magnitude as compared with their dimensions, the smallest of these magnitudes being almost infinitely large compared to the æthereal atom.

Chemical matter, which is relatively coarse grained, could not transmit indifferently disturbances of such vastly different wave-lengths. Our argument above that the velocity of transmission in solids is constant whatever the period of the swing, was based upon the assumption, which was stated, that the body was homogeneous. The body can only be regarded as homogeneous as long as the size of its molecules is very greatly less than the wave-length of the disturbance.

From the way that æther treats disturbances of very great and very small wave-length, we conclude therefore that, if æther have a corpuscular, or atomic, structure, it must be vastly finer in grain than the molecular structure of matter.

On the other hand, this evidence does not show that the structure of æther *is* grained—it might be *continuous*. This is a point which, as far as the writer is aware, cannot be decided on the present evidence. But things look as if the old metaphysical doctrine of a *plenum* were reviving, in the way metaphysical notions do.

The phenomena of electrostatics indicate that æther is in a state of shearing stress in the neighbourhood of electrically charged bodies. The oscillations set up by the discharge of a Leyden jar would, on this supposition, be due to the elastic recovery of the æther from this strain.

We may compare the condition of the neighbouring æther when a Leyden jar is charged to the condition of the part A of our iron wire (*vide ante*) when, having been grasped by the hand, it is twisted and maintained in this strained position by the grasp of the hand. The *discharge* of the Leyden jar is similar to the release of the grasp upon the end of the wire. In each case there is an elastic recovery of the strained substance setting up a vibratory motion which is transmitted as a wave.

Both electrostatic strain and wave oscillations take place in insulators, such as ebonite, which are not transparent to light. This shows that the presence of æther is not con-fined to transparent bodies.

The properties of a magnetic field may be represented by a whirling motion of æther.

An electric current is possibly produced by the sliding in opposite directions of portions of æther. This would come about if conductors (such as metals) are bodies penetrated by æther, but in which the æther loses its rigidity. Now, a body in which æther lost its rigidity could not transmit light, and in accordance with these conclusions is the striking fact that electric conductors are *opaque.*

More evidence is required, however, as to the nature of the difference between positive and negative electricity before electric and magnetic phenomena, other than radiations, can be adequately explained in terms of the properties of æther.

The periodic disturbances which we have so far found that æther transmits are *transverse* to the direction of transmission.

The question arises, *Does æther transmit also waves of longitudinal displacement?*

Every substance, whether solid, liquid, or a gas, resists compression and tends to spring back if compressed, and the compression is propagated throughout the substance at a speed which is constant for each substance. The disturbance is propagated as a wave, and the velocity of the wave depends upon the ratio of the *elasticity of volume* to the density. Elasticity of volume takes the place in waves of longitudinal displacement that rigidity (elasticity of shape) does in waves of transverse displacement.

No case of the propagation of a periodic disturbance of longitudinal displacement by æther has been certainly detected. One way, perhaps the simplest, of accounting for this would be the supposition that æther is incompressible. If a pressure produces no compression, there would only be an absolutely sudden transmission of pressure throughout the substance. Whether such *thrusts* constitute gravitation is not known, but Professor Lodge thinks it is possible that they do. A pull can always be resolved into a push, and if one portion of matter acts as a screen, warding off a part of the push of the æther from a neighbouring portion of matter, then the two portions of matter would tend to approach one another.

Lord Kelvin's suggestion as to a possible constitution of chemical matter, viz. that the chemical atoms are *vortices* of æther, would reduce all matter to one essential stuff—

æther. The mutual actions of different portions of ordinary or chemical matter, whether gravitational, electrical, or radiational, or chemical, would be due to action and reaction between vortexed and unvortexed æther.

The uncertainty which at present hangs round these speculations must not be allowed to prejudice the mind against accepting as scientific fact the existence of a vastly important kind of matter, which (except *possibly* in a vortexed state) does not appear to gravitate, although possessed of inertia; which is very rigid when proper means are taken to evoke its elasticity, but which does not stop the motion of gravitating matter; which is invisible, but without which there is no light.

CHAPTER XIV.

THE RÖNTGEN RAYS.

(Written March, 1896.)

WHEN an electric discharge from an induction coil is passed through a Hittorf or Crookes's tube containing a highly rarefied gas, a glow is seen in the tube, a well-known phenomenon of the electric discharge. Professor Röntgen has discovered that something else besides the visible glow proceeds from the tube—some agency which travels out in straight lines, or rays.

What kind or kinds of rays constitute this agency we do not know, and therefore the effects produced are said to be due to the "X rays," X standing for what has yet to be discovered, namely, the mode of motion of the rays.

Although we do not yet know what the rays are, we know a good deal about what they can do. They are able to produce three principal kinds of effect, by which we are made aware of their existence; first, they reduce the silver salt on a photographic dry plate; second, they make a fluorescent substance shine; third, they cause an electrical charge to leak away, however well insulated it may be.

When the vacuum tube through which a discharge is passing is wrapped up closely in black paper, or in a shield of black cardboard (which stops light and the ultra-violet rays of the spectrum), something (viz. X rays) passes through the black paper, and will reduce the silver salt on a photographic dry plate placed within a distance of, say, one or two yards. The experiment is, of course, carried out in a completely darkened room. If a piece of paper, painted

with barium platino-cyanide, be exposed at a similar distance, the paint fluoresces, glowing in the darkened
room.   Other materials, in addition to the black covering,
may be placed between the " vacuum-tube " and the screen
or the dry plate without stopping the X rays.   A book of a
thousand pages, two packs of cards, a one-inch deal board,
a still thicker sheet of ebonite, or a thin plate of aluminium,
each of these obstacles fails to quench the X rays, although
diminishing the intensity.   Every material yet tried is more
transparent to the X rays than some one of them is to each
of the other natural agencies (excepting gravitation) by
which effects are transmitted to a distance.

Thus we have already seen that the X rays pass through
black paper, which stops luminous rays and the ultra-violet
rays of the spectrum.   Electro-magnetic waves, such as
those set up in Hertz's experiments with electric oscillators,
will pass through a deal door, as X rays do; but electric
waves are stopped by a metal screen, which is by no means
opaque to X rays.   A sheet of copper, of silver, zinc, or lead
is unable to screen a needle from the attraction of a magnet,
but even a very thin sheet of iron does so quite effectually.
It does not appear from the published results that iron has
more effect than other metals in stopping X rays.

Although the X rays do not penetrate all qualities of
matter with equal facility, yet, so far as we have at present
learnt, the most noticeable factor in the dimming of the
rays is the actual amount of matter, as shown by the density
of the material.   Thus the X rays are more dimmed in
passing through a thin sheet of lead than through a similar
sheet of zinc, and platinum, which is much heavier than
lead, dims the rays still more.   Opacity to X rays is, however, not simply proportional to quantity of matter, though
much more nearly so than in the case of luminous rays.

When a body which is fairly opaque to the X rays is
placed between the covered vacuum-tube and the fluorescent
screen a shadow falls upon the screen—that is to say, the
parts of the screen which are behind the object either do

ENGLISH GRASS SNAKE.

(From X ray photograph taken by Newton & Co., 3, Fleet Street, E.C.)

not fluoresce at all, or only glow dimly, while the parts which are freely exposed to the X rays glow brightly. The shadow has the proper shape of the object to which it is due, which shows that the agency which emanates from the vacûum-tube radiates in straight lines, which justifies the use of the word "ray." If the hand be held between the vacuum-tube and the fluorescent screen the X rays pass almost freely through the fleshy parts, but are to a great extent quenched by the bones, and a dark shadow of the bony framework of the hand is thrown upon the screen, generally with a faint outer shadow of the fleshy covering. If a photographic dry plate is substituted for the fluorescent screen, a shadow-photograph is formed by the chemic action of the X rays around the parts of the plate which are screened by the bony framework of the hand. Any substance in the body which is less transparent to the X rays than the flesh is may be similarly detected, and this fact was successfully applied in the examination of bullet-wounds very soon after the first announcement of Professor Röntgen's discovery.

The action of the rays in discharging an electrified body furnishes, according to Professor J. J. Thomson, the most delicate test of their presence and the best means of obtaining relative measurements of their quantity. In Professor Thomson's experiments the vacuum bulb with the Ruhmkorff coil is placed inside a large packing-case covered with tin plate, but provided with a window which is closed only with a thin plate of aluminium or tinfoil. The tin-plated packing-case screens the outer room from all electric or luminous effects, the thin metal plate of the window allows the X rays to pass while assisting in the screening off of electric effects. When the X rays fall upon a highly electrified plate placed near the "window" the charge of electricity rapidly leaks away, as is shown by the movement of the spot of light thrown on a scale by the mirror of an electrometer to which the charged plate is connected. This discharge occurs whether the plate be positively or negatively electrified.

There does not appear to be any electric action upon a plate which is not charged before being exposed to the rays. The leakage from a charged plate occurs not only when the plate is surrounded by air, but also when it is embedded in a solid non-conductor, or dielectric, such as sulphur, paraffin wax, or ebonite, or a liquid non-conductor, such as paraffin oil.

Thus it appears, as far as experiment has gone, that the X rays have the power of penetrating every substance, and that every substance when exposed to their action becomes a conductor of electricity. So much for what the X rays can do.

With regard to the mode of motion of the X rays, we have already mentioned that they proceed in straight lines, and so far they resemble rays of light. But light rays, when passing from air into a denser material, move more slowly, and consequently when the ray of light enters the denser medium at a slanting angle, the direction of the ray is altered, just as when one end of a roller reaches the shallowing slopes of the seashore the wave-crest swings round and advances directly upon the beach. The X rays, on the contrary, are not bent out of their direct course when they pass from a rarer to a denser medium, and we conclude therefore that they travel at the same speed through all kinds of matter. This is indeed a remarkable property. The æther waves, both long and short, with which we have hitherto had to do, all travel at the same, or nearly the same, velocity in air or in a vacuum; but when traversing dense matter the case is different. Not only does light travel more slowly in glass than in air, but whereas in air or in a vacuum, the red rays, which have a longer wave-length, travel at the same rate as the shorter violet rays; in glass this is not the case, as we know from the fact that the violet rays are bent more out of their course, whereby we get the dispersion of light which gives the spectrum band. If, therefore, the X rays are due to a new kind of æther wave, it would seem that the wave-length must be such that all kinds of matter are, in comparison, of equally fine grain or equally coarse grain. As the different degrees of closeness of grain make no difference to the velocity, it would seem that the wave-length must be either very great or very small. The fact that the X rays excite fluorescence seems strongly against the supposition of a great wave-length, which would connote a slow vibration, for it is a rule that, when one vibration excites another, the second cannot be *quicker* than the first, though it may be slower. There *may* be exceptions, but the rule seems pretty general. The photographic action of the X rays likewise points to rapid vibration, if the X rays be waves of æther, for it is the shorter waves of light which are most potent in photography. The X rays traverse a powdered material as freely as when it is in coherent form, which indicates that they do not undergo reflection. In fact, they make their way through dense matter in any state, much as sound-waves traverse a dusty atmosphere, slightly quenched, if the number of particles be very great, but otherwise in-different to the dense particles strewn about in the trans-mitting medium.

If the X rays are due to waves of æther of extremely short wave-length, two suppositions are possible : first, that they are a superior ultra-violet light, very ultra indeed ; second, that the vibrations are not transverse, as in light, but longitudinal, as are the vibrations in sound-waves.

There is a good deal of difficulty in accepting the first supposition, on account of the extreme difference of magni-tude which would be necessary to account for so great a change of properties.

There are fewer objections to the view that the X rays are longitudinal vibrations, but confirmatory evidence must be obtained before the view can be considered more than a provisional supposition; for we have not as yet any other evidence that the æther vibrates longitudinally. Whether it can do so or not, depends upon whether it is com-pressible. We know that the rigidity of æther (its resistance

to a shearing-stress) is very great as compared with its
density, whence the great velocity of the transverse light-
waves. The resistance to compression is probably much
greater than the resistance to shearing, otherwise longitudinal waves (waves of compression) would presumably be
as much *en evidence* as the ordinary light-waves. The
tremor transmitted through the solid earth from a seismic
disturbance is a case in point. There the rocks transmit a
wave of distortion and a wave of compression, and the
latter is the more noticeable component of the earthquake
shock. If the compressibility of æther stood in the same
relation to its rigidity as is the case of the rocks of the
earth's crust, the formation of waves of compression, longitudinal waves, would presumably provide some striking
phenomena during every case of the reflection of light.
It seems likely, therefore, that the compressibility of the
æther is very small, but if it be not absolutely *nil*, longitudinal waves are possible. They would be of extremely
short wave-length and would travel immensely faster than
light. It has been suggested that the transmission of
electrostatic force is effected by means of longitudinal
waves of æther, but the velocity of this transmission has
never yet been measured, and in the mathematical treatment of electricity it is assumed to be infinite, *i.e.* the æther
is treated as being incompressible.

An alternative supposition remains, viz. that the X rays
are not due to æther vibrations of any sort, but to some
other mode of motion, such as that of flying particles of
matter electrically charged; or of a chain of particles transmitting a charge from one to another, *something* after the
fashion in which a row of marbles on the hollow rim of
a solitaire board transmits a shock from one end to the
other. Such flying particles, or such a chain of particles,
would act in many ways as a current of electricity, and
would presumably be deflected from its straight path by
a magnet. Such a deviation is observed in the case of the
rays which emanate from the negative pole, or cathode, in

a vacuum tube, but the X rays pursue their usual straight-forward course, unaffected by the magnet.

At present, therefore, we cannot say much more as to the mechanism of the transmission of X rays than this: that the longitudinal wave hypothesis appears to be a convenient one.*

* The theory of longitudinal vibrations has already fallen somewhat into disfavour. Sir G. Stokes is inclined " to regard the disturbances as non-periodic, though having certain features in common with a periodic disturbance of excessively high frequency " (*vide* " Nature," Sept. 3rd, 1896).

CHAPTER XV.

GREAT ASTRONOMERS.

"GREAT ASTRONOMERS," by Sir Robert Ball, is a pleasantly written book which will be found easy reading by the least learned.   The chapters on Hipparchus and Ptolemy merit careful reading, if only to correct the false impression of astronomical science which is given by the popular plan of commencing the subject with a statement of the Copernican system (followed by Kepler's laws and Newton's discovery of gravitation), and then plunging into the details of descriptive astronomy as developed with the aid of modern telescopes.   There is a good deal which should precede all this, whether the subject be studied historically in books, or practically by observation (with the unaided eye) as a part of natural history.   The task of the earliest astronomers was to observe, record, and represent the apparent positions and motions of the bodies in the sky, an indispensable preliminary to any rational theory of their relations to each other and to the earth.   The individual cannot do better than recapitulate in his own experience the historical development of the science, and begin the study of astronomy with actual observation of the position and motions of the sun and moon, stars and planets.   This might be followed by learning the "use of the globes" (an old-fashioned accomplishment now too much neglected), which, with a careful perusal of the late Sir George Airy's short course of "Lectures upon Astronomy," would give a far more satisfactory and more satisfying knowledge of astronomy than most people possess, in spite of the avidity with which the latest discoveries in the science are read and discussed.

Hipparchus, an astronomer of Rhodes, about 160 B.C. made as complete an inventory as possible of the fixed stars, and investigated the motions of the sun, moon, and planets. For the reduction of his observations he created trigonometry, as, in modern times, Newton created the differential calculus to solve the problems presented by gravitation. Both cases are excellent examples of the stimulus which the observation of the stars has given to pure mathematics—a stimulus similar to that which industrial pursuits have so often given to the development of chemistry and physics. Ptolemy, an Alexandrian astronomer, whose recorded observations extend from A.D. 127 to A.D. 151, realised the figure and position of the earth as a ball poised without support in space, and surrounded on all sides by stars. The constancy of the relative brightness of the stars as viewed from different localities showed him that their distance was great as compared with geographical distances. The constancy of the relative positions of the fixed stars, as recorded by successive generations of astronomers, indicated that in their apparent motion round the earth they moved all together, as if rigidly connected: they appeared to be set in a spherical shell, having the earth at its centre. Ptolemy perceived that the diurnal motion of the stars could be accounted for, geometrically, either by the existence of a revolving celestial sphere, or by the revolution of the earth. The former supposition implied a revolution of enormous velocity, since the dimensions of the celestial sphere were great compared with the dimensions of the earth; but there appeared to be insuperable difficulties in the way of accepting the alternative supposition that the earth itself revolved. The laws of motion, so simple in form yet so hard to grasp, were not at that time understood; the conception of an atmosphere possessing the same motion as the solid earth was beyond the scientific insight of the time; hence the doctrine of a revolving celestial sphere was preferred to that of a rotating

P

earth.   The absence of any shift in the position of the "fixed" stars seemed also to show that the earth had no motion of translation.   Ptolemy analysed with singular skill the apparently complex motions of the planets, and showed how these motions might appear as a combination of circular movements to a spectator placed above (or below) their plane of motion.

Copernicus, reviewing the Ptolemaic system fourteen hundred years later, showed the unsoundness of the old argument about the currents of air, and removed this principal objection to the hypothesis of the earth's rotation. He also showed how much simpler and less artificial the motions of the planets become if they and the earth are represented as revolving round the sun.   The telescopic observations of Galileo contributed to the subversion of the older astronomy; he greatly advanced our knowledge of the laws of motion, and his discoveries inaugurated modern descriptive astronomy.   Tycho continued the work of Hipparchus and of other cartographers, and provided, without telescopic aid, tables of astronomical positions and motions so full and accurate that from them Kepler was able to detect the small ellipticity of the planetary orbits, and to deduce the relation between the distances of the planets from the sun and the periods of their revolution. Sir Robert Ball makes an excellent suggestion for the curious in these matters—viz. to construct the old-fashioned "cross-stick" such as Tycho used in his earlier work, and with its aid to measure the angles between stars with a view to ascertaining the approach to accuracy attainable by such means of measurement.

Newton, having shown that the force of the earth's gravity, if diminished in proportion to the square of distance, would account for the observed deflection of the moon from a rectilinear path, proved further, with the aid of his specially invented calculus, that the laws of planetary motion discovered by Kepler can be accounted for by the existence of an attractive force, diminishing with the square of the distance, acting between the sun and the planets. Newton's contemporary, Halley, went to St. Helena in 1676—a precursor of Sir John Herschel—and mapped the southern stars. He also showed that certain comets revolve periodically round the sun, and are not, as had been supposed, merely casual visitors from external space. Flamstead mapped the fixed stars with improved instruments at the newly founded Observatory of Greenwich; and Bradley, about 1730, detected that minute displacement of the position of a star which is due to the fact that the velocity of the earth in its orbit is appreciable when compared even with the tremendous velocity with which light travels. The minuteness of this displacement (or "aberration") of the stars may be judged from the fact that the earth's velocity is only $\frac{1}{10000}$th part of the velocity of light. This discovery seems to mark the commencement of modern precision in astronomical measurement.

William Herschel was the pioneer of sidereal astronomy, the practical discoverer of nebulæ, and the first, we believe, to detect the existence of a gravitating force among the fixed stars, as shown by the existence of revolving double stars. Laplace carried on Newton's astronomical work, perfecting his calculus and extending the applications of the law of gravitation. He was also the author of the celebrated hypothesis of the evolution of the solar system through the mutual attractions of the parts of an immense nebulous mass. John Herschel completed his father's survey of the sky by observations at the Cape, and perfected the methods for determining the orbits of double stars. This latter work showed that gravitation among the fixed stars follows the same law as among the members of the solar system. In the life of Lord Rosse we note the personal interest of the narrator in the scientific work of a former patron; for Sir Robert Ball, when a young man, had charge of the great reflector at Parsonstown. It was with this instrument that the interesting spiral, or whirl, structure in nebulæ was discovered. Sir George Airy deserves to be ranked among the greatest of astronomers, although no one epoch-making discovery fixes his name in the mind of the unlearned. To his work we mainly owe the present high efficiency of Greenwich Observatory, so well maintained by his successor. He was a man distinguished for .clear thought, sound judgment, and practical ability. His researches on the moon's motion, on the transit of Venus, and on the density of the earth, are still fresh in the minds of astronomers. The account of Le Verrier consists mainly of the story of the discovery of the planet Neptune from the irregularities in the motion of Uranus, a crowning triumph to the fame of Newton. The last chapter in the book is devoted to the life of Adams, who shared with Le Verrier the credit of the discovery of Neptune. In common with Professor Newton, of Yale, Adams shares also the credit of showing that the November meteor-swarm revolves in a cometary orbit. Perhaps the most remarkable work of his later years was the revision of Laplace's calculations relating to the acceleration of the moon's motion. Adams showed that Laplace was mistaken in concluding that the ordinary application of the laws of gravitation between celestial bodies is sufficient to account for the shortening (*reckoned in days*) of the moon's revolutions round the earth. It seems probable that the true cause is the *lengthening* of the earth's day, owing to tidal friction. This is an excellent example of the importance of revising the work of even the most careful men of science. We recently had a conspicuous example of this in the discovery of Argon.

There is nothing in astronomical biography more picturesque than the life of the first Herschel, and of his devoted sister. The subject has often been treated, but never, we think, quite so successfully as in Miss Clerke's "The Herschels and Modern Astronomy." Miss Clerke brings out clearly the position occupied by William Herschel as the pioneer of *stellar* astronomy. He gave an impetus to the study of the descriptive part of astronomy such as it had not received since the time of Galileo. No such powerful impetus was again received until the last few years, in which astronomers have perfected the methods of taking photographs of long exposure, on which are impressed the images of myriads of stars, too faint to be seen directly with the telescope, but whose accumulated rays slowly produce a visible effect upon the photographic plate.

The eighteenth-century astronomers were fully occupied in working out the consequences of Newton's great generalisation (of gravitation) as exhibited in the complex motions of the members of the solar system. The stars were valuable to astronomers mainly as affording fixed points of reference by means of which the movements of planetary bodies were registered and recorded. The graduated circle, and other measuring appliances of astronomical instruments had been greatly improved; but for some time there had been no striking advance in the optical power of telescopes, and descriptive astronomy was accordingly at a discount.

William Herschel was fired with an ambition to explore the whole of the starry regions beyond the solar system. He realised more vividly and forcibly than his contemporaries that the stars were not placed on a distant background, all parts of which might be regarded as almost equally remote from our system ; but that the space-penetrating power of the telescope could reveal stars so distant that, in comparison, the brighter stars would be our near neighbours. He hoped to discover an orderly arrangement and grouping of the stars throughout the length and depth of space, perhaps even to place the sun (which was to him but a star) among the constellations. The measure of success which he attained in his ambitious project was largely due to his perseverance in two principles which he always kept before him—viz. to get as much *light* as possible, and never to lose an hour of a clear starlit night. Even the meetings of the Royal Society were neglected except when the moon was near the full.

Herschel was a splendid workman, and toiled with extraordinary diligence at grinding specula—an art in which he was unrivalled, but which he does not seem to have reduced to a science. His twenty-foot reflector, with a speculum of twenty inches, was enormously more powerful than any telescope which had previously been made, and the forty-foot reflector, of about four feet aperture, was, when first set up, as powerful and fine an instrument as any, except a very few, of the greatest telescopes of the present day. Unfortunately, the metal specula preserved their perfection of polish and of figure for a very short time, and constant renewal of the surface was required—an arduous and difficult task. The failing powers of old age prevented him from renewing the speculum of the forty-foot telescope, as he had been wont to do with the twenty-foot—a fact which probably accounts for the disappointing performance of the great telescope after it had been in use for only one or two years.

The *nebulæ* may almost be considered as Herschel's discovery, so little was known of them before his time ; and he recognised the salient fact that they are grouped in greater numbers where the stars thin out, towards the poles of the Milky Way. The capital importance of the Milky Way itself in the scheme of star-distribution was brought out clearly by his "gauges" in and near that extraordinary region. His observations of double stars, of which next to nothing was known before his time, showed, as we have said, that there was a gravitating or attractive force acting not only between the members of the solar system, but also between star and star.

The life of Sir John Herschel is, in many respects, a striking contrast to that of his father. Sir William was a splendid amateur, who took up astronomy in middle age. His son was a highly trained scientific man, who at one-and-twenty knew all that Cambridge could teach. But his life does not make such good reading as his father's ; it lacks the variety of picturesque incident ; and we miss the unchecked enthusiasm of the old pioneer, who made splendid discoveries, or splendid mistakes, almost every day; who once nearly lost his life in his impatience to try a new speculum before the scaffolding which he mounted was properly fixed.

The first important astronomical work of John Herschel was the revision of his father's observations, particularly of double stars and nebulæ. The work on double stars showed that the gravitating force between stars discovered by the elder Herschel followed the well-known law of gravitation as established for the members of the solar system. This revision completed, John Herschel, in 1833, set out for the Cape, with a twenty-foot reflector, to complete his father's survey of stars and nebulæ by a corresponding survey of the southern hemisphere of the sky. Four years of unremitting labour sufficed for the actual observations, though the reduction and publication of the results took twice as long. We can only here refer to his generalisation on the subject of the Milky Way, which he had now studied throughout its whole extent. The plane of the Milky Way, he says, "is to sidereal, what the ecliptic is to planetary, astronomy, a plane of ultimate reference, the ground plan of the sidereal system." The "Cape observations" having been completed, John Herschel ceased to work as an observing astronomer. He was only forty-six when he returned from the Cape; but his father had married late, and already the observations of the two Herschels covered a period of nearly seventy years. The days of pioneering surveys of the sky were now over, and the younger Herschel was probably acting in the best interests of astronomy in following those studies which led to the development of spectrum analysis and photography, to both of which he materially contributed. Having long been the best all-round man of science of his day, Sir John Herschel died in 1871 (a century after the time when his father first attempted the manufacture of a telescope), leaving a third generation of astronomers to bear the honoured name of Herschel.

CHAPTER XVI.

NATURAL PHILOSOPHY.

THE natural philosophers of old times had probably as much ability and industry as their successors, but they did not understand the rules of the game, as their successors do—to some extent.

They read the book of Nature upside down. They were determined to wrest what they wanted from Nature, instead of being content to find out what is, and to adapt their wants accordingly.

Man was the centre of the Universe, and external Nature was expected to conform to his *à priori* views of what was right and proper. The processes of Nature were to be so modified by man's ingenuity that the curse of work and the curse of death should be revoked.

The failure of such false science has led to a better appreciation of the limitation of our powers, and of the directions in which they may profitably be exercised.

To almost every branch of modern science, there has been a perverted counterpart.

Magic, as far as it dealt with material things, was a false dynamics.

Astronomy was doubly false, being based on metaphysical notions of the necessity for motion in the "perfect" curve, and being further perverted to the purposes of astrology.

Alchemy was a false chemistry, and the belief in "perpetual motion" was part of a false science in physics.

Natural history, even in recent times, has been perverted

by the supposed regulation of the living world, in accordance with our own notions of what is just and right.

Mechanical magic may be defined as the attempt to produce mechanical effect by mental effort. The magician could produce action without calling out the correlative reaction. He could produce physical effect " at a distance," without a material connection with the body acted upon. The belief in art magic served to secure a property in mechanical inventions, which is now secured by patent rights. It was futile to imitate an invention which worked solely through the personal power of the magician.

Witchcraft, as far as it consisted in attempts to influence the mind, was less irrational than mechanical magic. The action of mind upon mind is, at all events, not unthinkable. This branch of witchcraft still flourishes, but we call it " telepathy."

Astrology, which professed to trace a connection between celestial motions and the course of human life, was not *inherently* absurd. True, there is no apparent physical mechanism to connect the occurrences, but neither is there in the case of the connection between sun-spots and terrestrial magnetism. Astrology became absurd when men obstinately refused to draw the logical conclusion from the data accumulated during centuries of observation, that the observed coincidences of the horoscope were not systematic, but casual.

Astronomy itself, in the times immediately preceding Copernicus, Galileo, and Kepler, was hopelessly encumbered with metaphysical notions. The circle was the perfect curve, *therefore* the planets moved in circles. The sun was the perfect light, *therefore* there were no spots on the sun ; and so on. Perhaps astronomers were more truly scientific in the earlier days, in Chaldæa, and in Egypt. Nature-worship can scarcely have perverted natural knowledge so hopelessly as did metaphysics.

The transmutation of metals was not intrinsically absurd as a guiding idea for experiment. The absurdity was in the persistency of the belief in transmutation, after centuries of work, during which the evidence gained was mostly on the opposite side. The history of alchemy provides one of the most humiliating examples of the capacity of the mind to continue believing what it wishes to be true.

The search for the *elixir vitæ* was certainly contrary to common sense, which is not quite the same as saying that it was wholly irrational; but it had less semblance of justification than the search for the philosopher's stone.

The later endeavour of the perpetual-motionist to discover a self-acting motor, was an attempt to escape from the curse of work, as the search for the *elixir vitæ* was an attempt to escape from the curse of death. It would be too much to say that the idea of a self-acting motor was intrinsically absurd, but undoubtedly the *belief* in the possibility of such an arrangement was quite unscientific. Modern research upon the conservation of energy, and its corollary, the degradation of energy, have pretty well put an end to the perpetual-motionists.

The six-days geology, which was formerly the general interpretation of the Book of Genesis, is not to be classed among the false sciences. It was a case of evidence; on the one hand an authoritative book, on the other an ill-digested mass of observations of rocks and fossils. The geologic observations were, in course of time, extended and arranged, and now it is recognised that "the Bible was not written to teach science."

On the other hand, a good deal of natural theology was very false science indeed. Its methods were comparable to those of "circular" astronomy. Everything had to be made to fit our private notions of fitness. The chief difference is that the notions were in one case geometrical, and in the other case moral.

Providential arrangements (such as the existence of raptorial species, for the purpose of sparing sick animals the misery of prolonged suffering) are no longer a fashionable interpretation of natural history. In their place we have the battle for existence, "which rages through all time and in every field, and its rule is to give no quarter—to despatch the maimed, to overtake the halt, and to trip up the blind " (Martineau).

We are only now beginning to recognise that the survival of the "fittest" only means, as some one has said, the survival of the "fightingest." The perversion of the phrase, " survival of the fittest," has threatened a false science of sociology. The doctrine that in human society we are to let the struggle for existence have free play, trusting that the " fittest" will survive, has come upon us as a Nemesis. We used to interpret the animal world by moral notions, and now we are recommended to revert from altruism to animalism.

The laws of conservation and the doctrine of evolution provide the material for the greater part of the natural philosophy of to-day.

The two great laws of conservation, the law of conservation of mass and the law of conservation of energy, have familiarised the world with the doctrine " that in Nature nothing is lost." The form of matter changes, but the quantity remains unaltered. The form of energy changes, but its "mechanical equivalent," that is to say, the amount of mechanical effect it would produce, if all the energy were converted into the motion of a mass of matter, remains constant.

In Nature nothing is lost ! Surely no dictum of science was ever more contrary to common sense, the sense of what is reasonable, as derived from the ordinary experience of life.

Let us take first the law of conservation of mass. It has been the common experience of mankind for ages that solid fuel burns away, producing some smoke, and leaving a little ash behind, no other visible substance being formed in the burning. There is nothing else produced which is either visible, tangible, or, as far as common experience goes, ponderable. Common experience failed to afford an insight into the changes in the forms of matter which happen during burning.

Lavoisier's *expériences* corrected common sense in this, and we now say that in chemical change no matter is destroyed. Common experience tells us, however, that things have a tendency to wear out and get used up. Iron rusts, and soil loses its fertility. Later chemical research has confirmed this common experience, and shows that the general tendency of chemical changes is to the production of material less capable of reacting chemically, and less available for the work of the world.

The carbon chemically combined in plants is still capable of burning, and, as a constituent of vegetable food, is converted by animals into carbonic acid. Certainly no carbon has been annihilated, but when it has got to the final form of carbonic acid, there is little more that carbon can *do*. If it is not lost, it is at all events pretty securely locked up. Were it not for the extraneous energy the earth receives from the sun's rays, the carbon in carbonic acid would be, in this sense, irretrievably lost.

In the processes of industrial chemistry, we may often see the manufacturer's art applied to the production of some material in a form in which it is chemically active ; we may cite, for instance, iron-smelting and alkali-making. Dwellers in the neighbourhood of Middlesborough, or of Widnes, realise better than most people that for every ton of iron a far greater quantity of an effete slag is formed—slag which has accumulated in masses greater than the pyramids of Egypt—and that for every ton of soda there are produced nearly two tons of alkali waste.

The balance of chemical change is on the side of diminished chemical activity. Nature grows weary, and protests afterwards against changes which she does not prevent, and often scarcely hindered at the time.

It is, however, in the domain of physics rather than in chemistry, in transformations of energy more than in transformations of matter, that Nature's protest against change is most strikingly shown. We do not refer merely to the property of inertia, in virtue of which all bodies *resist* a change in their state of rest or motion. In this case, if the resistance be overcome and the motion of the body be increased, energy is stored up in the increased momentum of the body; but when such exchanges of energy are examined in detail, there always appears a residual phenomenon which mars the simplicity and completeness of the result.

Take the case of any of the well-known mechanical appliances. A small force is made to raise a large weight, and if we multiply the force by the distance through which its point of application is moved, the product would be equal to that of the weight multiplied by the distance which the weight is raised, were it not for the inevitable friction which fritters away a part of the mechanical energy in the less available form of heat. It is true that the heat so produced has its mechanical equivalent. Heat can be converted into mechanical work, and one unit of heat converted into mechanical energy produces 1370 foot-pounds, the same amount of mechanical work which was required to produce the unit quantity of heat; but if we are dealing with the unit quantity of heat, it is found in practice to be impossible to transform the whole of it into mechanical work; part still remains in the form of heat. Perhaps one half of the unit quantity of heat may be converted into 685 foot-pounds, leaving the other half still in the form of heat energy. This is an example of the conservation of heat energy, but it is a somewhat unsatisfactory form of conservatism, for the half unit of heat which is left is heat at a lower temperature, or heat-level, than that with which we started, and is less available for further transformation. Only a reduced proportion of this heat can be converted into the higher or more available forms of energy.

Hirn's experiments with the steam engine illustrate this point. The loss of heat between the boiler and condenser is accounted for by the mechanical work done, and the mechanical equivalent of heat determined in this way is identical with the heat equivalent of work determined by Joule's method of warming a liquid by the friction of paddle-wheels driven by a falling weight. But the steam engine cannot be worked without a transference of heat between the boiler and the condenser, over and above the heat that is used to overcome resistance. This surplus heat is degraded from the high temperature of the boiler to the low temperature of the condenser. It is still the same in quantity, but it is less available for transformation, less ready to undergo further change.

In this case of the running down or degradation of energy an external mechanical effect is produced, and the result is in accordance with our everyday experience, and therefore is consistent with the expectations of common sense. We cannot work without growing tired, and it does not seem unreasonable that something analogous to fatigue should be shown by inorganic nature. But the inorganic world shows loss of vigour, not only after effort, but after any change, even though not accompanied by external effect. It is in such phenomena of diminished vigour, resulting from all those natural changes which proceed without effort and without the accomplishment of work, that we best see Nature's protest against all change.

One of the best illustrations, which is not only a striking one, but has been carefully and quantitatively examined, is afforded by the expansion of compressed air into a vacuum. In Joule's well-known experiments, a vessel containing compressed air and provided with a stop-cock was attached to a similar vessel which had been exhausted of air. The two were placed in the same vessel of water, and the temperature carefully noted. The stop-cock was then opened, and the compressed air was allowed to rush into the empty vessel. The empty vessel is thereby heated, and the full vessel cooled; but on stirring the water of the outer or containing vessel so as to equalise the temperature through-out, it is found that the temperature of the system of bodies has, as a whole, undergone no alteration. The only change is that the energy of the air is less available than it was before. There is the same amount of energy *in* it, but in order to get energy *out* of it we should have to expend work upon it, *e.g.* by compressing it again into a smaller space.

A similar and even more important case is that of the equalisation of temperatures which is constantly taking place by the flow of heat from hotter to colder bodies, or from hotter to colder parts of the same body. The amount of heat received by the cold body is equal to that lost by the hot body; but whereas work, mechanical, electrical, chemical, and so forth, could always be got out of the two bodies as long as there was a difference of temperature between them, now that the temperatures are the same no work or effect can be obtained from the interaction of the two.

The tendency of all forms of energy to transform themselves into heat, and the tendency of heat-energy to become uniformly distributed, and thus ineffective, is one of the most important modern generalisations from the study of physical forces. A levelling process appears to go on everywhere in the inorganic world, and if the tendencies, which are so distinctly seen to operate now, are part of a continuing order of things, then we cannot avoid the logical conclusion that the world tends towards a state of death in life in which all the mass and all the energy of the present cosmos are undiminished but impotent. Then there shall be no more change.

Such is the dreary prospect afforded by the attempt to push beyond the limits of our experience the conclusions to which observation of inorganic nature undoubtedly leads.

What a different prospect is unfolded by the doctrines which have grown out of the scientific study of the animated world! Natural selection, the survival of the fittest, evolution—familiar terms expressing the generalisations of biological science—point to differentiation, development, and progress. It is easy to observe the effect of the doctrines of evolution and development in the optimistic tone as to progress and the future of human society adopted by many writers of late years.

How is it, one may well ask, that the tendency of things appears so different when looked at from the point of view of the physical and of the biological sciences? Perhaps it is that the progress of species, in which the exercise of volition plays an important part, resembles the processes of the industrial arts, in which the finer and more service-able forms of matter are produced by aid of thought and contrivance, rather than the ordinary operations of inorganic nature. We have pointed out how the manufacture of metallic iron involves the simultaneous production of greater quantities of the effete iron slag. Something analogous seems to be indicated in the frightful waste of animal life, and in the extinction of species. Death, however, removes these from the sphere of action—a difference between the organic and the inorganic world, the importance of which will be realised more particularly by those who have studied the processes of chemical change.

Neither should it be forgotten that the student of physical science deals statistically with the phenomena he investi-gates, whereas the work of the student of biological science, and the student of mankind also, is to a large extent con-cerned with individuals. Physical science deals with the properties of energy as exhibited by matter. The units or individuals of matter are atoms and molecules, and we cannot examine these individually. Could we do so we might, and probably should, find that the history of any one of the small number of molecules which we might individually study would differ from the history of the body of which they form a part, as much as the fortunes of the individual man may differ from the general lot of the human race. Recurring to the example of the compressed air expanding into a vacuous vessel, it might happen that certain molecules would receive impacts so directed and so timed that their velocity of motion and their individual energy would be greatly increased. If chance directed our attention to such cases, we might be led to suppose that the change which accompanied the expansion of air had been "progressive" in its character, since it acted for the "benefit" of the "fortunate" molecules. But the statistical study of the phenomenon as a whole would show that, in spite of the "development" of individuals, there had been a general lowering of vigour all round.

However these things may be, it is undoubtedly the fact that the powers of the animated world are ultimately derived from the inorganic source of physical energy, and sooner or later the powers of organic development must cease if the phenomena of degradation of energy as exhibited by the inorganic world are really universal in their application.

INDEX.

A

B

C

D

THE END.

LONDON : PRINTED BY WILLIAM CLOWES AND SONS, LIMITED,
STAMFORD STREET AND CHARING CROSS.

www.ingramcontent.com/pod-product-compliance
Lightning Source LLC
Chambersburg PA
CBHW030818020726
47499CB00006B/1976